DERAILED EVOLUTION

DERAILED EVOLUTION

Nils K. Oeijord

iUniverse, Inc.
New York Lincoln Shanghai

DERAILED EVOLUTION

Copyright © 2005 by Nils K. Oeijord

iUniverse books may be ordered through booksellers or by contacting:

iUniverse
2021 Pine Lake Road, Suite 100
Lincoln, NE 68512
www.iuniverse.com
1-800-Authors (1-800-288-4677)

ISBN-13: 978-0-595-37353-6 (pbk)
ISBN-13: 978-0-595-81750-4 (ebk)
ISBN-10: 0-595-37353-4 (pbk)
ISBN-10: 0-595-81750-5 (ebk)

Printed in the United States of America

Contents

1. Preface

All DNA is continuously in close contact with manmade and highly reactive and mutagenic/carcinogenic chemicals as well as manmade and highly reactive and mutagenic/carcinogenic radiation. This situation is undoubtedly a general genetic catastrophe of the World. Also, this situation means that human evolution is derailed.

"We [the human race] do not have much time to prove that we are not the product of a lethal mutation."

Science 263: 181, 1994

"I almost think it is the ultimate destiny of science to exterminate the human race."

Thomas Love Peacock

2. The General Genetic Catastrophe of the World

A general genetic catastrophe has been developing during the last 120 years, but mostly after 1950.

The HIGH and INCREASING prevalence and HIGH and INCREASING incidence of genetic damage/genetic disorders/genetic diseases all over the world logically prove that we have a general genetic catastrophe (GGC).

Science-based technology has polluted the whole World with some 1 000 000 mutagenic chemicals in addition to mutagenic radiation. Even gasoline vapor and diagnostic use of X-rays permanently damage our chromosomes and genes, and burning cigarettes produce smoke containing 800 mutagens. Chemical by-products of our scientific civilization, like Dioxines, PCBs, and PAHs, are destroying DNA even in the Arctic

Tragically, the scientific community is misunderstanding or defining away the GGC. When will Science wake up? The History of Science tells us that it may take time. Science has perhaps never understood fundamentally NEW situations?

3. Genetic Catastrophe Means Derailed Evolution

The genetic catastrophe consists of four major genetic "epidemics"—those of cancer, vascular disease, musculoskeletal disease, and behavioral disease. There are identified approximately 250 common genetic diseases, and approximately 7,000 "rare" genetic diseases.

The list of genetic damage is growing daily. The natural rate at which mutations and genetic damage occur is reasonably constant. The larger the gene the more common genetic diseases of the gene are. Hence, more common genetic diseases are caused by mutations/damage to very large genes.

Genetic damage is far more common than generally understood. There is insufficient recognition of the magnitude of the genetic catastrophe. The concept of genetic disease has expanded during the last twenty years. The concept has now expanded to be virtually all encompassing. Even infectious diseases have some relationships to our genes. Nobel-laureate Paul Berg said "all human disease is genetic."

The natural mutation frequency is about 1 mutation per 100,000 genes per generation. This value proves that present-day pollution is an absolute genetic tragedy because this value lies near the critical value that separates evolution from extinction.

The strong increase in the gene damage of genetic diseases like cancer, vascular diseases, Alzheimer's, Parkinson's, diabetes, osteoporosis, obesity, etc, is perhaps the best illustration of the magnitude of the genetic catastrophe. Certainly the victims of genetic damage are the victims of modern human activities (including science and science based technology).

More than 80 percent of us are tortured and finally killed by cancer, vascular diseases, and other non-rare genetic diseases. Most of the rest of the population are tortured, and often finally killed, by so-called rare diseases. Science, in general, does not speak about a genetic catastrophe. On the contrary, by using words like syndrome, disorder, disease, illness, defect, deficiency, failure, etc, instead of genetic damage, science, in general, is defining away and covering up the genetic catastrophe.

In the US, a rare disease is defined as "one that afflicts no more than 200,000 people." However, one out of 10 Americans suffers from just such a rare disease. And more than 6,000 diseases have been classified as rare, according to the US National Organization for Rare Disorders. If (when) these 6,000 diseases reach, on avarage, the 200,000 people level, then each American, on average, will suffer from more than four rare diseases. Most rare diseases are genetic diseases caused by gene damage.

In our globally polluted world our genes suffer damage just through day-to-day living. Our DNA is constantly under attack from mutagenic chemicals and radiation causing a steady build-up of heritable gene damage, in spite of the best efforts of our DNA repair enzymes and other repair mechanisms, such as the SOS repair mechanism, the cell-suicide mechanism, sterility, infertility, natural and unnatural abortion, recurrent miscarriages, death at an early age, and natural and unnatural selecton.

There is no evidence that there is a dose below which there is not a mutagenic effect. A small dose of chemicals/radiation to a large population does more harm to the gene pool than a large dose to a small population. In fact, the chemicals causing the genetic catastrophe currently often meet all environmental standards, and often these chemicals are present at non-detectible levels. We don't know what's actually happening at the genetic level. Besides, genetic damage is particularly insidious because it can take several generations for the effects to show up. Moreover, the total exposure to mutagenic chemicals and radiation is unknown.

The blood of Inuit people and polar bears contains a large number of mutagenic chemicals in high concentrations. Certainly, the romantic notions of "wilderness", "health", "science", and "technology" are outdated.

Although the individual is often powerless to avoid exposure to widely used chemicals, there are examples of mutagens to which people voluntarily expose themselves, for example cigarette smoke.

Unfortunately, mutagenic chemicals occur in widely divergent chemical groups, ranging from simple compounds such as formaldehyde to complex ones such as alkaloids. With every breath of cigarette smoke, the body is confronted by more than 800 mutagenic chemicals which include dioxin-like compounds.

Apropos of dioxins, the gene damage of dioxins is absolutely devilish: The DNA double-helix acts like a zipper, which can open and close. A dioxin molecule acts like a thread-like object which is preventing the zipper from being closed!! All mother's milk (human and non-human) of the world contains dioxins, now and in the future!!

Some 150,000 (an increasing number) American infants are born annually with serious birth defects, which include brain anomalities and cleft palate. These 150,000 infants are probably born in our avarage environment. "44 million Americans can't read" because of the genetic damage of dyslexia. An individual may be superintelligent but he/she is unable to learn to read because of a specific genetic damage. These 44 million (an increasing number) individuals are probably born in, and live in, our average environment. The genetic damage of cancer kills approximately 30 percent of the total US population, while the genetic damage of cardiovascular diseases kills more than 30 percent of the US population. A 20-year-old individual may suddenly die of a cardiovascular disease.

Because every human individual has about 35,000 functional genes, there is an endless number of possible heritable genetic diseases. Furthermore, there is an endless number of possible chromosome anomalies such as chromosome breaks. Ionizing radiation is extremely efficient at causing chromosome breaks. Even gasoline vapor and high-voltage fields cause chromosome breaks. These messy breaks are difficult or impossible for cells to repair correctly. Evidence shows that here the cell's repair-system is fallible even when it is confronted only by a minimal challenge.

Even relatively low (diagnostic) doses of X-rays cause gene damage. Data strongly suggest that preconceptional exposure of the mother to diagnostic doses of X-rays increases the risk of offspring with Down's syndrome. New research results show that diagnostic doses of X-rays are both mutagenic and carcinogenic.

Chemicals in the general pollution damage crucial DNA repair genes. Besides, heavy metals from background pollution damage repair enzymes (change their form and function). When a repair-gene is damaged, this gene damage will magnify the consequences of the cell's subsequent exposures to all mutagens (chemicals and radiation), because of the cell's diminished ability to repair gene damage correctly. And free radicals and other mutagens attack our DNA all the time.

Cells somehow sense gene damage and activate a suicide program, called apoptosis, to kill themselves so that gene damage is not perpetuated. This is the ultimate "brake" against cancer as well. The p53-gene is known to trigger apoptosis in response to gene damage in a cell. If the p53-gene is damaged, then an important way of preventing gene damage from being passed on to the next generation is lost. How can we stop the gene/genetic damage explosion?

Genetic diseases appear to be identical across species. The elimination of genetic diseases in animals can only be accomplished through selective breeding in a clean environment. This is not a theory, this is a fact. However, as we have shown above, the situation for our genes is enormously difficult. It looks as if the genetic diseases will be the winner of the race.

Today twice as many dies of diabetes (per 100,000 people) as before insulin was introduced. Today, in general, twice as many dies of a genetic disease (per 100,000 people) as before a medicine was introduced.

The genetic damage to human sperm cells is so enormous that we can recognize directly, by looking through a light microscope, that about 50 percent of the cells are abnormal. There is perhaps nothing that can prevent all sperm cells to become abnormal. After all, normal cells live in the same polluted environment as the abnormal cells. Of course, the genetic damage to egg cells is even more dramatic because they are relatively few in number.

Men born with a birth defect have a doubled risk, compared with other fathers, of having a child with a birth defect, a large population study revealed. Note that this situation, according to simple mathematics, leads to a genetic damage explosion. The genetic catastrophe of breast cancer develops extremely fast: Of all the women with breast cancer, only a tenth have family histories of the disease, and half of this group has a heritable gene damage causing breast cancer!

Department of Education estmates that 20 percent of Americans are learning-disabled. This result is obviously wrong, because the dyslexics (20 percent of Americans) are not the only learning-disabled people in the US. The number of 6,000 for rare diseases in the US is also wrong. Behind each of the 6,000 diseases (mostly genetic diseases) there are lots of different types of gene damage.

In the general population in Rochester, N.Y., 3 percent had, in the year 2000, the genetic damage of Tourette's syndrome, and 20 percent had the (genetic?) damage of a tic disorder. The rate of 3 percent in the general population is about 50 to 75 times higher than typical estimates.

It has been estimated that more than 50,000 chemicals are in common use in the United States. Most of these chemicals have not been tested for mutagenicity. Of those that have been tested for mutagenicity, about 20 percent produced mutations in the Ames test. Of the several million chemicals in "uncommon" use in the world, perhaps as much as about 1,000,000 are causing gene damage (mutations). These chemicals are working in your cells at this very moment.

The natural (evolutionary) mutation rate is about one mutation per fertilized egg. If the mutation rate seriously exceeds one mutation per fertilized egg (it obviously does), then evolution is derailed. Also, if the mutation rate is one, or slightly less than one, mutation per fertilized egg, and, if, at the same time, natural selection is seriously disturbed (it is), then evolution is derailed.

Behavior is a biochemical event. Behavioral disorders often have a genetic basis. Most neurobehavioral syndromes are genetic damage. With regard to survival of the human species, damage to genes for behavioral traits ("behavioral genes") is much more dangerous than damage to genes for physical traits. A large and rapidly increasing percentage of the population suffers from genetic damage of behavioral disorders such as learning disabilities, depression, antisocial personality disorder, Tourette's syndrome, schizophrenia, bipolar disorder, eating disorders, attention deficit disorder, hyperactivity, fragile-X syndrome, autism, Lesch-Nyhan syndrome, Down syndrome, etc. If (when) only 100 genetic behavioral disorders reach, on average, the 1 percent level, then humanity is destroyed?

4. A Short History of the Genetic Catastrophe

In the Stone Age the inventions of the stone axe, fire, and the burning of woodlands created the first man-made mutagenic pollution: smoke. Of course, the invention of the saw (4000 BC) increased the production of mutagenic smoke.

Fired brick (4th century BC-) meant more gene-damaging smoke. The energy-providing use of natural gas and oil (4th century BC-) added to mutagenic pollution. Gas warfare (China 4th century BC-) was not good for our genes.

Salting and smoking (techniques of food preservation) invented by the Chinese, of course, put a lot of mutagens into our food. (Even NaCl is mutagenic.)

Gunpowder (9th century-) produced a large amount of gene-damaging smoke. War technology is one of the greatest producers of mutagenic smoke. Rockets (1241-), the gunpowder blowpipe (1304-), the gunpowder cannon (1320s-), the musket (1570-), the revolver (1630-), the modern gun (1807-), the repeter rifle (1860-), the modern war machines (20th century-), the atomic bomb (1945-), were new stages in the genetic catastrophe.

The politicians and the industry did not want to develop the atomic bomb, but the leading scientists worked hard to persuade them that the atomic bomb was important for the war. This is a good example of how science rules the World.

Cigar and cigarette smoking (Europe 1518-) enlarged the genetic catastrophe, putting several hundred mutagens into our body.

Public lighting (street oil lamps, 1700-) polluted our streets with mutagens and carcinogens. The widespread use of coal (1800-) certainly was a genetic tragedy. Matches ('safety' matches 1855-) directly and indirectly represented additional "progress".

Serious pollution by heavy metals (heavy metals are mutagenic) started in the Copper Age, and increased in the Bronze Age (bronze = copper + tin) and the Iron Age. Metal money (8th century BC-) meant serious mutagenic heavy metal pollution. Metal pipes made out of lead (3rd century BC-) meant serious mutagenic lead pollution. The use of porcelain (1st century-) meant increased mutagenic heavy metal pollution of our body and our environment. Color printing (9th century-) spread more gene-damaging heavy metals in the environment (including our body). The invention of the mercury thermometer (1672) was a mutagenic tragedy.

Canned food (1815-) and certain chemical pesticides (1800-) put more mutagenic heavy metals in our food and our bodies. Non-synthetic chemical pesticides (1800s-) such as creosote and lead arsenide were bad for our DNA. In more modern times aluminium (1850s-) directly, but mostly indirectly, was a great "improvement" of our scientific civilization. Needless to say, the genetic catastrophe is a by-product of our scientific civilization. Production and use of superphosphates (fertilizer) (1900-) pollutes our food with mutagenic and carcinogenic cadmium.

Modern synthetic pesticides (1950-) were a direct genetic tragedy. Today more and more of these terrible manmade molecules are shown to be mutagenic and carcinogenic.

The Industrial Revolution (in the 1700s and 1800s) increased the production of numerous gene-damaging chemical substances. All new inventions caused increased industrial pollution and increased mutagenic/carcinogenic pollution. The inventions of modern technology
(1850-) created new ways of destroying our genes and chromosomes.

Here is a list of some of the gene-damaging inventions of the Industrial Revolution:

Asphalting roads (France 1730-) (Asphalt contains mutagenic/carcinogenic pollution.)
The steam engine (1769-)
Machine tool (1775-)
Chlorination and ozonization (1700s-)

Lead bullet (1700s-)
Metal cartridges (1799-)
Metal-coated bullet (1800s)
The locomotive (1813-)
Calculating machine (commercial) (1820-)
Public railway (1825-)
Steam turbine (commercial use, 1831-)
High-explosive shell (1837-)
Stamping (sheet-metal manufacture) (1838-)
Vulcanization of rubber (1839-)
Metal hulls and propeller propulsion (1843-)
Metal construction (1848-)
Armour-plating of warships (1850s-)
Cannon with grooved bore (1858-)
Photography (1858-)
Drilling tower (in the West) (1859-)
The London Underground (1863-)
(Despite the gene-damaging sulphurous gases emitted by the system, the London Underground was immediately successful. It transported some 10 million passengers in its first year of service in 1863. The London Underground was electrified in 1890. Note that electromagnetic radiation breaks our chromosomes.)
Transcontinental railroad (1869-)
Modern bicycle (1874-)
Standardization and scientific management (1880-)
The electrical transformer (1880s-)
Gasoline-driven automobile (1885-)
The electric generator (1890s–)
Fibreglass (1893-)

The electrical transformer and the electric generator produce electromagnetic radiation breaking our chromosomes.

The 20th century was, of course, a global genetic catastrophe. Millions of persistent manmade mutagenic chemicals are now found absolutely everywhere in the environments of the World, including our bodies. We now know that the persistent organochlorines damage our genes. The cities of the World are a genetic Hell filled with polluting cars and chimneys, and electromagnetic smog. All kinds of things that can damage our DNA.

Probably as much as 90 percent of known inventions causing mutagenic pollution (including mutagenic radiation) date from the 20th century. Below is a list of some of the gene-damaging (directly or indirectly) inventions of the 20th century:

Gas turbines (1900-)
Low-cost cars (1908-)
The New York metro (1912-)
Electrification of the railways (20th century)
Modern ships (20th century)
Modern submarines (20th century)
(The first military use of a submarine dates back to 1775, to the American War of Independence.)
Synthetic glues (1930-)
(The old adhesives were non-mutagenic and based on plaster, resin, starch, fish glue, casein, gelatine, wax, eggwhite, etc)
Nuclear energy (electric power) (1951-)
(In the "Radiation Area" of our nuclear industry the workers' chromosomes are being damaged. Hundreds of manmade radioactive isotopes have polluted the whole World. Science explains away this fundamental genetic catastrophe.)
Modern medicines (20th century-)
(Many modern chemical medicines are known to be gene-damaging. Pharmacopoea Danica 1933 tells us that creosote and mercury are good medicines! Today diagnostic use of X-rays damages our chromosomes. Never ever trust science!)
The compact cassette (1962-), the modern computer (1968-), the color home video recorder (1972-), etc, etc meant enormous mutagenic/carcinogenic pollution. A burning PC, say, produces extremely damaging pollution. Even a non-burning PC or a non-burning TV emit mutagenic/carcinogenic chemicals.

Directly and indirectly, the products and by-products of the Electric Age (1879–1946), the Electronic Age (1947–1972), and the Information Age (1973-) represent a total genetic catastrophe of the World.

In 1962 an important book was published: *Silent Spring* by Rachel Carson. Science had created and produced enormous amounts of extremely damaging pesticides. The author Carson made the World aware of this global catastrophe. She was persecuted because science told us that she was wrong. But the environmentalists and the politicians reacted to a certain extent. In 1963, the first "Clean Air

Act" was passed to control air pollution, and in the 1960s environmentalism became a political movement. However, the general genetic catastrophe of the World has exploded since the 1960s. Science answered Carson by inventing an enormous number of new dangerous and persistent pesticides and other kinds of mutagenic environmental poisons.

The 1972 book *A Blueprint for Survival* (written by a large number of scientists) claimed that "social disintegration is a major cause of mental disease." Most other claims in this well known book are wrong as well, history has shown us, and this book systematically avoided the issue of gene/genetic damage. Unbelievable but true: Science did—and still does—systematically neglect, avoid, boycott, misunderstand, explain away, or cover up the fundamental problem of gene-damaging pollution and gene damage.

The oceans are an ultimate accumulation site of the chemical and radioactive products and by-products of our scientific civilization. Even the arctic is now totally polluted by mutagenic and carcinogenic chemicals. The polar bears contain more mutagenic environmental poison than many types of dangerous industrial waste. Arctic seagulls filled with mutagenic pollution are falling dead to the ground. Also the arctic fishes are poisoned by mutagenic chemicals. Science neglects this fact. Never trust science at the dining table.

The effects of mutagens are not separable, but instead additive or mutually reinforcing. Rapid cumulative genetic collapse of our scientific civilization is now a real threat. But science is silent. Politicians, labor leaders, industry leaders, and science fight together for the creation of new polluting industry based on new science-based technology and new science-created mutagenic chemicals. Their decisions are always scientific, and their advisers, consultants, and experts are the scientists. No one listen to "unscientific" people. Remember that scientific management was established in the 1880s, and note that fact is: the general genetic catastrophe exploded in the 1880s.

As a matter of fact, science has created the genetic catastrophe. How? By replacing non-mutagenic natural products by mutagenic manmade synthetic products. Examples: Plastics replaced wood products. Detergents replaced soaps made from natural fats. Synthetic fibres replaced natural fibres. Chemical fertilizers replaced organic manure. Fossil fuels replaced wood. Nuclear energy replaced fossil fuels. Etc. Etc. Nature does not produce a single molecule for which there is not an enzyme capable of breaking it down, in order to perpetuate the cycle of life, growth, death, and decay. Synthetic products cannot normally be broken down in this way. Therefore, even the slightest amount of synthetic products constitutes serious pollution, because these chemicals find their way into our bodies, where they attack our genes, chromosomes, cells, and organs. The activities of scientists and industrial man are having a very serious effect on our bodies and our society. "Even in the realms of science, take nothing at face value." (John Waller: *Fabulous Science*.) "The myth of scientific detachment." (John Waller: *Fabulous Science*.)

We must understand the past on its own terms without any reference to "what happened next." A new generation of leading historians of science has shown that seemingly crucial scientific experiments were often fatally flawed and that results were often modified to suit the case being argued. Evidence is now available to show that many leading scientists used political influence to advance their cause. "The great biologist Louis Pasteur suppressed data that didn't support the case he was making. Albert Einstein's theory of general relativity was only 'confirmed' in 1919 because an eminent British scientist massaged his figures. Joseph Lister's famously spotless hospital wards were actually notoriously dirty. Gregor Mendel, supposed father of the science of heredity, never grasped the fundamental principles of 'Mendelian' genetics." (From John Waller: *Fabulous Science*.) Science is a field filled with propaganda and fundamentalism, that is Science represents progress and Science represents the only Truth. And Science is a God. Even John Waller writes about "serious sins against science."

Gregor Mendel's epoch-making paper was first presented in 1865, but was discovered and understood only in the early 1900s. Mendel never grasped the basic tenets of Mendelian genetics. And Darwin ignored Francis Galton's ingenious particulate theory of heredity (available in the 1860s), but "his" theory of blending inheritance was to cause him immense discomfort. The word "gene" was not coined until 1903. Chance mutations were first studied in the 1910s. In the 1930s it was fully realized that Darwinism and Mendelism made a fine matching pair. But even today—70 years later—science does not understand that evolutionary theories are crucial to the future of the human populations. This sad situation partly explains why science still does not understand that we have a genetic catastrophe. Of course, scientific products, by-products, activities, and results damaged our genes even before 1903, but before 1903 science did not know the word "gene damage". This fact does illustrate how extremely dangerous the totality of science is. It does show that science is blind when it comes to the real big things. Karl Popper made a useful distinction between *discovery* and *verification* in the development of knowledge. Science only becomes reliable knowledge, Popper argued in 1959, after its validity has been tested over the course of many years. Popper was "inclined to think that scientific discovery is impossible without faith in ideas which are of a purely speculative kind, and sometimes even quite hazy." Certainly, polluting the whole World (including our bodies) with mutagenic and carcinogenic chemicals and radiation is quite hazy.

5. How to Explain Away the Genetic Catastrophe

The following list contains 20 ways to explain away the genetic catastrophe and derailed evolution.

1. "Business as usual. All these genetic diseases have always existed."

2. "The upward trend is not real. This is due to better diagnosis."

3. "These diseases are caused by our lifestyle."

4. "These diseases are caused by unhealthy food."

5. "These diseases are caused by viruses."

6. "These conditions are environmental."

7. "The increased lifespan disproves that we have a genetic catastrophe."

8. "The upward trend is due to increased lifespan."

9. "A genetic catastrophe is not possible because of natural selection."

10. "A genetic catastrophe is not possible because all gene-damaged fetuses are naturally aborted."

    [It's true that 75 percent of all fetuses are naturally aborted.]

11. "A genetic catastrophe is not possible because the immune system kills all gene-damaged cells."

12. "A genetic catastrophe is not possible because all gene-damaged cells commit suicide (apoptosis)."

13. "A genetic catastrophe is not possible because of DNA repair."

14. "These conditions are a result of evolution."

15. "These conditions are adaptations."

16. "The explanation is that our bodies simply wear out."

17. "No, you don't have any evidence."

18. "You [Nils K. Oeijord] are simply a crackpot with an odd, paranoid 'genetic catastrophe'

    fixation who deserves little of our time." (Eric)

19. "The GC is an absurd idea caused by paranoia. It is utter garbage, unsupported by evidence, unreferenced, cranky." (Alec)

20. "The scientific community publishes a very nice journal called "Human Mutation". We get it at our lab. It contains many very thorough and detailed papers regarding all aspects of human genetic diseases. It also publishes the Mutation and Polymorphism Report. Tons of good information in there as well. Our Centers for Disease Control publishes a weekly Morbidity and Mortality Report. And you're trying to tell me that with the thousands of well trained and concerned people around the world reading these publications every week—not to mention the thousands of other good, relevant scientific journals—that somehow ALL of them are completely missing some absolutely catastrophic trend? (Eric)

Author's comments:

Dear Eric!

Thanks a lot, Eric! You seem to have documented above that I'm the discoverer of the genetic catastrophe and derailed evolution.

Best regards,

Nils K. Oeijord

6. Informative Correspondence

Correspondence 1

Letters: Question and Answer
Question:

Nils K. Oeijord
Styrmannsv 62
9014 Tromso
2002-04-21
Norway

Forskningsavdelingen
Statistisk sentralbyrå
Postboks 8131 Dep.
0033 Oslo

Genetic damage of genetic disorders, facts and information

Could you kindly "print out" the following information on each of the several thousand genetic disorders of the human population of Norway, and send it to me:

Year the genetic disorder/genetic disease/genetic damage was first detected (in Norway)
Number of new cases per year
Total number of cases

Also, could you kindly send me a copy of the prognoses of the Statistisk sentralbyrå for each of the genetic disorders for the next 100 years.

Please answer in English because your answer is extremely important and will be used in my next book on the genetic catastrophe. If I've to pay for this service, it's OK.

Most sincerely,
Nils K. Oeijord

Answer:

Statistisk sentralbyrå
Statistics Norway

Nils K. Oeijord
Styrmannsv. 62
9014 Tromso

Oslo, 21 June 2002
Your ref.:, Our ref.: 02/954-1
Executive officer: Anne Mundal
Division for Health Statistics

Thank you for your request about information on genetic disorders. Unfortunately we don't have the information that you ask for, but we recommend that you contact Norwegian Institute of Public Health, P.O. Box 4404 Nydalen, N-0403 OSLO, telephone 22 04 22 00.

Best regards,
Anne Mundal

Correspondence 2

Letters: Question and Answer
Question:

Nils K. Oeijord
Styrmannsveien 62
9014 Tromso
2002-08-05
Norway

Telephone/fax	+47 776 71 215
Mobile phone	+47 99 10 13 86
E-mail	n-oeij@frisurf.no

Norwegian Institute of Public Health
P.O. Box 4404 Nydalen, N-0403 OSLO
Norway

Genetic damage of genetic disorders, facts and information

Could you kindly "print out" the following information on each of the several thousand genetic disorders of the human population of Norway, and send it to me:

Year the genetic disorder/genetic disease/genetic damage was first detected (in Norway)
Number of new cases per year
Total number of cases

Also, could you kindly send me a copy of the prognoses for each of the genetic disorders for the next 100 years.

Please answer in English because your answer is extremely important and will be used in my next book on the genetic catastrophe. If I've to pay for this service, it's OK.

Very sincerely,
Nils K. Oeijord

Answer:

Folkehelseinstituttet

Nils K. Oeijord
Styrmannsv 62
9014 Tromso

Dear Nils K. Oeijord

We have received your letter sent August 5 2002. Unfortunately, there is no registration centrally of cases of genetic diseases, so we have no way of answering your question. As far as we know there is no indication that the frequency of genetic diseases is increasing in Norway.

Sincerely yours,

Per Magnus, MD
Head of Department

Correspondence 3

Letters: Question and Answer
Question:

Nils K. Oeijord
Styrmannsveien 62
9014 Tromso
2002-08-10
Norway

Telephone/fax	+47 776 71 215
Mobile phone	+47 99 10 13 86
E-mail	n-oeij@frisurf.no

Norwegian Institute of Public Health
P.O. Box 4404 Nydalen, N-0403 OSLO
Norway

Non-central registration of genetic diseases/genetic disorders/genetic damage in Norway.

Thank you very much for your answer of August 6 2002 to my letter sent August 5 2002. Where can I find non-central registration of genetic diseases/genetic disorders/genetic damage in Norway?

Very sincerely,
Nils K. Oeijord

Answer:

Folkehelseinstituttet

Nils K. Oeijord
Styrmannsv 62
9014 Tromso

September 11, 2002

Dear Nils K. Oeijord

Thank you for your letter sent August 10 2002. Patients with genetic diseases will some times attend the Departments of Medical Genetics at our four Universities. Additionally, they will seek help from general practitioners. However, I doubt whether they can be of much help in giving numbers that will help in judging the occurrence of genetic disorders in Norway.

Sincerely yours,

Per Magnus, MD
Head of Department

Correspondence 4

Letters: Question and Answer
Question:

Nils K. Oeijord
Styrmannsveien 62
9014 Tromso
2002-09-16
Norway

Telephone/fax	+47 776 71 215
Mobile phone	+47 99 10 13 86
E-mail	n-oeij@frisurf.no

Department of Medical Genetics
University of Oslo
P. O. Box 1130 Blindern
0318 Oslo

Genetic damage of genetic disorders, facts and information

Could you kindly "print out" the following information on each of the several thousand genetic disorders of the human population of Norway, and send it to me:

Year the genetic disorder/genetic disease/genetic damage was first detected (in Norway)
Number of new cases per year
Total number of cases

Also, could you kindly send me a copy of the prognoses for each of the genetic disorders for the next 100 years.

Please answer in English because your answer is extremely important and will be used in my next book on the genetic catastrophe. If I've to pay for this service, it's OK.

Very sincerely,
Nils K. Oeijord

Answer (E-mail!!):

[September 19, 2002]

Re your letter of 2002-09-16
Unfortunately the information you ask for is not available in Norway (or in any other country, I believe).

Sincerely yours,

Carl Birger van der Hagen
Prosektor, overlege
Institutt for medisinsk genetikk
Universitetet i Oslo
Boks 1036
0315 Oslo
tlf +47 22 11 98 81

Correspondence 5

Letters: Question and Answer
Question:

Nils K. Oeijord
Styrmannsveien 62
9014 Tromso
2002-09-16
Norway

Telephone/fax	+47 776 71 215
Mobile phone	+47 99 10 13 86
E-mail	n-oeij@frisurf.no

Department of Medical Genetics
University of Bergen
Jonas Liesvei 65
5021 Bergen

Genetic damage of genetic disorders, facts and information

Could you kindly "print out" the following information on each of the several thousand genetic disorders of the human population of Norway, and send it to me:

Year the genetic disorder/genetic disease/genetic damage was first detected (in Norway)
Number of new cases per year
Total number of cases

Also, could you kindly send me a copy of the prognoses for each of the genetic disorders for the next 100 years.

Please answer in English because your answer is extremely important and will be used in my next book on the genetic catastrophe. If I've to pay for this service, it's OK.

Very sincerely,
Nils K. Oeijord

Answer:

Haukeland University Hospital
Center for medical genetics and
molecular medicine

Mr. Nils K. Oeijord
Styrmannsveien 62
9014 Tromso
Bergen, 17. September 2002

Genetic damage of genetic disorders, facts and information

In response to your inquiry received here today, I'm sorry to inform you that no such data exist. Moreover, there are no ways to generate such data in a reliable way, regardless of resources put into such an undertaking. We lack data even for relatively "common" genetic disorders, as there is no mandatory registration of (genetic) disease in Norway (apart from cancer).

Our Surgeon General (Helsedirektoeren) recommended in the 1950'ies that some registration of selected disorders (considered inherited at that time) should take place at the Institute of Medical Genetics, University of Oslo. Thus, there may still be some registrations of inherited disease in the archives of this institute. This registration was never properly funded, was not compulsory and was completely abandoned many years ago. The head of the Institute of Medical Genetics at the time this registration was initiated was professor Jan Mohr (appointed later on to the Chair in medical genetics, University of Copenhagen). The head of the registry ("Arvelighetsregistret") throughout most of the registration period was professor Kaare Berg, recently retired.

Sincerely,
Helge Boman, M.D.
Professor of Medicine (Medical Genetics)
University of Bergen

Correspondence 6

Letters: Question and Answer
Question:

Nils K. Oeijord
Styrmannsveien 62
9014 Tromso
2002-09-16
Norway

Telephone/fax	+47 776 71 215
Mobile phone	+47 99 10 13 86
E-mail	n-oeij@frisurf.no

Department of Medical Genetics
University of Tromso
9037 Tromso

Genetic damage of genetic disorders, facts and information

Could you kindly "print out" the following information on each of the several thousand genetic disorders of the human population of Norway, and send it to me:

Year the genetic disorder/genetic disease/genetic damage was first detected (in Norway)

Number of new cases per year
Total number of cases

Also, could you kindly send me a copy of the prognoses for each of the genetic disorders for the next 100 years.

Please answer in English because your answer is extremely important and will be used in my next book on the genetic catastrophe. If I've to pay for this service, it's OK.

Very sincerely,
Nils K. Oeijord

Answer:

They did answer in Norwegian!! Moreover, the answer was worthless. First, they even used the telephone to avoid writing a letter!!

Correspondence 7

Letters: Question and Answer
Question:

Nils K. Oeijord
Styrmannsveien 62
9014 Tromso
2002-09-16
Norway

Telephone/fax	+47 776 71 215
Mobile phone	+47 99 10 13 86
E-mail	n-oeij@frisurf.no

Department of Medical Genetics
Norwegian University of Science and Technology (NTNU)
7491 Trondheim

Genetic damage of genetic disorders, facts and information

Could you kindly "print out" the following information on each of the several thousand genetic disorders of the human population of Norway, and send it to me:

Year the genetic disorder/genetic disease/genetic damage was first detected (in Norway)
Number of new cases per year
Total number of cases

Also, could you kindly send me a copy of the prognoses for each of the genetic disorders for the next 100 years.

Please answer in English because your answer is extremely important and will be used in my next book on the genetic catastrophe. If I've to pay for this service, it's OK.

Very sincerely,
Nils K. Oeijord

Answer:

They did not answer!!

7. Derailed Evolution Examples: A List of Genetic Damage in Humans

48 XXYY syndrome, genetic damage of

49 XXXXX syndrome, genetic damage of

49 XXXXY syndrome, genetic damage of

5-Nucleotidase syndrome, genetc damage of

5-Alpha-reductase 2 deficiency, genetic damage of

5-Oxoprolinase deficiency, genetic damage of

6-Pyruvoyl-tetrahydropterin synthase deficiency, genetic damage of

7-Dehydrocholesterol reductase deficiency, genetic damage of

A

Aagenaes syndrome, genetic damage of

Aarskog Ose Pande syndrome, genetic damage of

Aase Smith syndrome, genetic damage of

Aase syndrome, genetic damage of

ABCD syndrome, genetic damage of

Abdallat Davis Farrage syndrome, genetic damage of

Abdominal aortic aneurysm, genetic damage of

Abdominal cystic lymphangioma, genetic damage of

Abdominal defects, genetic damage of

Abdominal musculature, absent microphthalmia joint laxity, genetic damage of

Abdominal neoplasms, genetic damage of

Aberrant subclavian artery, genetic damage of

Ablepharon macrostomia syndrome, genetic damage of

Ablutophobia, genetic damage of (possible)

Abnormal systemic venous return, genetic damage of

Abruzzo Erickson syndrome, genetic damage of

Absent corpus callosum cataract immunodeficiency, genetic damage of

Absent T lymphocytes, genetic damage of

Acalvaria, genetic damage of

Acanthocytosis, genetic damage of

Acanthocytosis chorea, genetic damage of

Acanthosis nigricans muscle cramps acral enlargement, genetic damage of

Acarophobia, genetic damage of (possible)

Acatalasemia, genetic damage of

Accessory pancreas, genetic damage of

Acetyl-CoA alpha-glucosaminide-N-acetyl transferase deficiency, genetic damage of

Achalasia, genetic damage of

Achalasia alacrimia syndrome, genetic damage of

Achalasia microcephaly, genetic damage of

Achalasia, familial esophageal, genetic damage of

Achalasia-Addisonianism-Alacrimia syndrome, genetic damage of

Achard syndrome, genetic damage of

Achard-Thiers syndrome, genetic damage of

Acheiropodia, genetic damage of

Achondrogenesis, genetic damage of

Achondrogenesis Kozlowski type, genetic damage of

Achondrogenesis type 1A, genetic damage of

Achondrogenesis type 1B, genetic damage of

Achondrogenesis type 2, genetic damage of

Achondroplasia, genetic damage of

Achondroplasia Swiss type agammaglobulinemia, genetic damage of

Achondroplastic dwarfism, genetic damage of

Achromatopsia, genetic damage of

Achromatopsia incomplete, X-linked, genetic damage of

Acid maltase deficiency, genetic damage of

Acidemia, isovaleric, genetic damage of

Acidemia, propionic, genetic damage of

Ackerman syndrome, genetic damage of

Acne rosacea, genetic damage of

Acoustic neurofibromatosis, genetic damage of

Acoustic neuroma, genetic damage of

Acoustic schwannomas, genetic damage of

Acousticophobia, genetic damage of (possible)

Acquired autoimmune hemolytic anemia, genetic damage of (possible)

Acquired hypoprothrombinemia, genetic damage of (possible)

Acquired ichthyosis, genetic damage of (possible (possible)

Acquired prothrombin deficiency, genetic damage of (possible)

Acral dysostosis dyserythropoiesis, genetic damage of

Acral renal mandibular syndrome, genetic damage of

Acro coxo mesomelic dysplasia, genetic damage of

Acro fronto facio nasal dysostosis, genetic damage of

Acrocallosal syndrome, Schinzel type, genetic damage of

Acrocephalopolydactyly, genetic damage of

Acrocephalosyndactyly Jackson Weiss type, genetic damage of

Acrocephaly pulmonary stenosis mental retardation, genetic damage of

Acrocraniofacial dysostosis, genetic damage of

Acrodermatitis, genetic damage of (possible)

Acrodermatitis enteropathica, genetic damage of

Acrodysostosis, genetic damage of

Acrodysplasia scoliosis, genetic damage of

Acrofacial dysostosis ambiguous genitalia, genetic damage of

Acrofacial dysostosis atypical postaxial, genetic damage of

Acrofacial dysostosis Catania form, genetic damage of

Acrofacial dysostosis Preis type, genetic damage of

Acrofacial dysostosis Rodriguez type, genetic damage of

Acrofacial dysostosis Weyers type, genetic damage of

Acrofacial dysostosis, Nager type, genetic damage of

Acrofacial dysostosis, Palagonia type, genetic damage of

Acrokeratoelastoidosis of Costa, genetic damage of

Acromegaloid changes cutis verticis gyrata corneal, genetic damage of

Acromegaloid facial appearance syndrome, genetic damage of

Acromegaloid hypertrichosis syndrome, genetic damage of

Acromegaly, genetic damage of

Acromesomelic dysplasia, genetic damage of

Acromesomelic dysplasia Brahimi Bacha type, genetic damage of

Acromesomelic dysplasia Campailla Martinelli type, genetic damage of

Acromesomelic dysplasia Hunter Thompson type, genetic damage of

Acromesomelic dysplasia, Maroteaux type, genetic damage of

Acromicric dysplasia, genetic damage of

Acroosteolysis dominant type, genetic damage of

Acroosteolysis neurogenic, genetic damage of

Acroosteolysis osteoporosis skull and mandible changes, genetic damage of

Acropectorenal field defect, genetic damage of

Acropectorovertebral dysplasia, genetic damage of

Acrophobia, genetic damage of (possible)

Acropigmentation of Dohi, genetic damage of

Acrorenal field defect ectodermal dysplasia diabetes, genetic damage of

Acrorenal syndrome recessive, genetic damage of

Acrorenoocular syndrome, genetic damage of

Acrospiroma, genetic damage of

ACTH deficiency, genetic damage of

ACTH resistance, genetic damage of

Activated protein C resistance, genetic damage of

Acutane embryopathy, genetic damage of

Acute articular rheumatism, genetic damage of

Acute erythroblastic leukemia, genetic damage of

Acute febrile neutrophilic dermatosis, genetic damage of

Acute idiopathic polyneuritis, genetic damage of

Acute intermittent porphyria, genetic damage of

Acute lymphoblastic leukemia, genetic damage of

Acute lymphoblastic leukemia congenital sporadic aniridia, genetic damage of

Acute lymphocytic leukemia, genetic damage of

Acute megakaryoblastic leukemia, genetic damage of

Acute monoblastic leukemia, genetic damage of

Acute myeloblastic leukemia type 1, genetic damage of

Acute myeloblastic leukemia type 2, genetic damage of

Acute myeloblastic leukemia type 3, genetic damage of

Acute myeloblastic leukemia type 4, genetic damage of

Acute myeloblastic leukemia type 5, genetic damage of

Acute myeloblastic leukemia type 6, genetic damage of

Acute myeloblastic leukemia type 7, genetic damage of

Acute myeloblastic leukemia with maturation, genetic damage of

Acute myeloblastic leukemia without maturation, genetic damage of

Acute myelocytic leukemia, genetic damage of

Acute myelogenous leukemia, genetic damage of

Acute myeloid leukemia (generic term), genetic damage of

Acute myeloid leukemia, secondary, genetic damage of

Acute myelomonocytic leukemia, genetic damage of

Acute non-lymphoblastic leukemia (generic term), genetic damage of

Acute posterior multifocal placoid pigment epitheliopathy, genetic damage of

Acute promyelocytic leukemia, genetic damage of

Acute renal failure, genetic damage of

Acyl-CoA dehydrogenase, medium chain, deficiency of, genetic damage of

Acyl-CoA dehydrogenase, short chain, deficiency of, genetic damage of

Acyl-CoA dehydrogenase, very long chain, deficiency of, genetic damage of

Acyl-CoA oxidase deficiency, genetic damage of

Adactylia unilateral dominant, genetic damage of

Adam complex familial, genetic damage of

Adams Nance syndrome, genetic damage of

Adams-Oliver syndrome, genetic damage of

Addison's disease, genetic damage of

Adducted thumb syndrome recessive form, genetic damage of

Adducted thumbs Dundar type, genetic damage of

Adenine phosphoribosyltransferase deficiency, genetic damage of

Adenocarcinoid tumor, genetic damage of

Adenocarcinoma, genetic damage of

Adenoid cystic carcinoma, genetic damage of

Adenoma, genetic damage of

Adenoma of the adrenal gland, genetic damage of

Adenomelablastoma, genetic damage of

Adenomyosis, genetic damage of

Adenosine deaminase deficiency, genetic damage of

Adenosine monophosphate deaminase deficiency, genetic damage of

Adenosine triphosphatase deficiency, anemia due to, genetic damage of

Adenylosuccinate lyase deficiency, genetic damage of

Adie syndrome, genetic damage of

Adiposa dolorosa, genetic damage of

Adolescent benign focal crisis, genetic damage of

Adrenal adenoma, familial, genetic damage of

Adrenal cancer, genetic damage of

Adrenal disorder, genetic damage of

Adrenal gland hyperfunction, genetic damage of

Adrenal gland hypofunction, genetic damage of

Adrenal hyperplasia, congenital, genetic damage of

Adrenal hypertension, genetic damage of

Adrenal hypoplasia congenital, X linked, genetic damage of

Adrenal incidentaloma, genetic damage of

Adrenal insufficiency, genetic damage of

Adrenal macropolyadenomatosis, genetic damage of

Adrenal medulla neoplasm, genetic damage of

Adrenocortical carcinoma, genetic damage of

Adrenogenital syndrome, genetic damage of

Adrenoleukodystrophy, genetic damage of

Adrenoleukodystrophy, autosomal, neonatal form, genetic damage of

Adrenoleukodystrophy, X linked, genetic damage of

Adrenomyeloneuropathy, genetic damage of

Adrenomyodystrophy, genetic damage of

Adult onset Still's disease, genetic damage of

Adult spinal muscular atrophy, genetic damage of

Adult syndrome, genetic damage of

Aerophobia, genetic damage of (possible)

Afibrinogenemia, genetic damage of

Aganglionosis, total intestinal, genetic damage of

Aganthia holoprosencephaly situs inversus, genetic damage of

Aggressive fibromatosis, genetic damage of

Aglossia adactylia, genetic damage of

Agnosia, primary visual, genetic damage of

Agonadism dextrocardia diaphragmatic hernia, genetic damage of

Agonadism mental retardation delayed bone age, genetic damage of

Agoraphobia, genetic damage of (possible)

Agrizoophobia, genetic damage of (possible)

Agyria pachygyria polymicrogyria, genetic damage of

Agyria-pachygyria type 1, genetic damage of

Agyrophobia, genetic damage of (possible)

AHD, genetic damage of

Ahumada-Del Castillo syndrome, genetic damage of

Aicardi syndrome, genetic damage of

Aicardi-Goutieres syndrome, genetic damage of

Aichmophobia, genetic damage of (possible)

Ailurophobia, genetic damage of (possible)

Akaba Hayasaka syndrome, genetic damage of

Akesson syndrome, genetic damage of

Aksu Stckhausen syndrome, genetic damage of

Al Awadi Teebi Farag syndrome, genetic damage of

Al Frayh Facharzt Haque syndrome, genetic damage of

Al Gazali Al Talabani syndrome, genetic damage of

Al Gazali Aziz Salem syndrome, genetic damage of

Al Gazali Donnai Muller syndrome, genetic damage of

Al Gazali Hirschsprung syndrome, genetic damage of

Al Gazali Khidr Prem Chandran syndrome, genetic damage of

Al Gazali Sabrinathan Nair syndrome, genetic damage of

Alagille-Watson syndrome, genetic damage of

Alar nasal cartilages coloboma of telecanthus, genetic damage of

Albers-Schonberg disease, genetic damage of

Albinism, genetic damage of

Albinism deafness syndrome, genetic damage of

Albinism immunodeficiency, genetic damage of

Albinism ocular, genetic damage of

Albinism ocular late onset sensorineural deafness, genetic damage of

Albinism oculocutaneous, Hermansky-Pudlak type, genetic damage of

Albinism, yellow mutant type, genetic damage of

Albinoidism, genetic damage of

Albrecht Schneider Belmont syndrome, genetic damage of

Albright like syndrome, genetic damage of

Albright Turner Morgani syndrome, genetic damage of

Albright's hereditary osteodystrophy, genetic damage of

Albuminurophobia, genetic damage of (possible)

Aldolase A deficiency, genetic damage of

Alektorophobia, genetic damage of (possible)

Aleukemic leukemia cutis, genetic damage of

Alexander's disease, genetic damage of

Alkaptonuria, genetic damage of

Allain Babin Demarquez syndrome, genetic damage of

Allan Herndon syndrome, genetic damage of

Allanson Pantzar McLeod syndrome, genetic damage of

Allergic angiitis, genetic damage of

Allergic autoimmune thyroiditis, genetic damage of

Allergic encephalomyelitis, genetic damage of

Allgrove syndrome, genetic damage of

Alliumphobia, genetic damage of (possible)

Allodoxaphobia, genetic damage of (possible)

Aloi Tomasini Isaia syndrome, genetic damage of

Alopecia, genetic damage of

Alopecia anosmia deafness hypogonadism syndrome, genetic damage of

Alopecia areata, genetic damage of

Alopecia congenita keratosis palmoplantaris, genetic damage of

Alopecia contractures dwarfism mental retardation, genetic damage of

Alopecia epilepsy oligophrenia syndrome of Moynaha, genetic damage of

Alopecia epilepsy pyorrhea mental subnormality, genetic damage of

Alopecia hypogonadism extrapyramidal disorder, genetic damage of

Alopecia immunodeficiency, genetic damage of

Alopecia macular degeneration growth retardation, genetic damage of

Alopecia mental retardation hypogonadism, genetic damage of

Alopecia mental retardation syndrome, genetic damage of

Alopecia totalis, genetic damage of

Alopecia universalis, genetic damage of

Alopecia universalis onychodystrophy vitiligo, genetic damage of

Alpers disease, genetic damage of

Alpha 1-antitrypsin deficiency, genetic damage of

Alpha-2 deficient collagen disease, genetic damage of

Alpha-ketoglutarate dehydrogenase deficiency, genetic damage of

Alpha-L-fucosidase deficiency, genetic damage of

Alpha-L-iduronidase deficiency, genetic damage of

Alpha-mannosidosis, genetic damage of

Alpha-sarcoglycanopathy, genetic damage of

Alpha-thalassemia, genetic damage of

Alpha-thalassemia-abnormal morphogenesis, genetic damage of

Alport syndrome, genetic damage of

Alport syndrome macrothrombocytopenia, genetic damage of

Alport syndrome, dominant type, genetic damage of

Alport syndrome, recessive type, genetic damage of

Alstrom's syndrome, genetic damage of

Alternating hemiplegia, genetic damage of

Alternating hemiplegia of childhood, genetic damage of

Alveolar hypoventilation syndrome, genetic damage of

Alveolar soft part sarcoma, genetic damage of

Alveolitis, extrinsic allergic, genetic damage of

Alves Dos Santos Castello syndrome, genetic damage of

Alzheimer disease, familial, genetic damage of

Alzheimer's disease, genetic damage of

Amathophobia, genetic damage of (possible)

Amaurosis congenita of Leber, genetic damage of

Amaurosis congenita of Leber, type 1, genetic damage of

Amaurosis congenita of Leber, type 2, genetic damage of

Amaurosis hypertrichosis, genetic damage of

Amaxophobia, genetic damage of (possible)

Ambral syndrome, genetic damage of

Ambras syndrome, genetic damage of

Ambulophobia, genetic damage of (possible)

AMC, genetic damage of

Amegakaryocytic thrombocytopenia, genetic damage of

Amelia cleft lip palate hydrocephalus iris coloboma, genetic damage of

Amelia facial dysmorphism, genetic damage of

Amelia X linked, genetic damage of

Amelo cerebro hypohidrotic syndrome, genetic damage of

Amelogenesis imperfecta, genetic damage of

Amelogenesis imperfecta local hypoplastic form, genetic damage of

Amelogenesis imperfecta nephrocalcinosis, genetic damage of

Ameloonychohypohidrotic syndrome, genetic damage of

Amenorrhea, primary, genetic damage of

Aminoacidopathies, genetic damage of

Aminoaciduria, genetic damage of

Aminopterin like syndrome without aminopterin, genetic damage of

Amniotic bands, genetic damage of

Ampola syndrome, genetic damage of

Amychophobia, genetic damage of (possible)

Amylo-1,6-glucosidase deficiency, genetic damage of

Amyloid angiopathy, genetic damage of

Amyloid polyneuropathy, transthyretin related, genetic damage of

Amyloidosis, genetic damage of

Amyloidosis of gingiva and conjunctiva mental retardation, genetic damage of

Amylopectinosis, genetic damage of

Amyoplasia, genetic damage of

Amyoplasia mandibulofacial dysostosis, genetic damage of

Amyotonia congenita, genetic damage of

Amyotrophic lateral sclerosis, genetic damage of

Amyotrophy fat tissue anomaly, genetic damage of

Anablephobia, genetic damage of (possible)

Anaphylaxis, genetic damage of

Anaplastic large cell lymphoma, genetic damage of

Anaplastic thyroid cancer, genetic damage of

Andermann syndrome, genetic damage of

Andersen's disease, genetic damage of

Andre syndrome, genetic damage of

Androgen insensitivity syndrome, genetic damage of

Androphobia, genetic damage of (possible)

Anemia, genetic damage of

Anemia sideroblastic spinocerebellar ataxia, genetic damage of

Anemia, pernicious, genetic damage of

Anemia, sideroblastic, genetic damage of

Anemophobia, genetic damage of (possible)

Anencephalus, genetic damage of

Anencephaly, genetic damage of

Anencephaly spina bifida X linked, genetic damage of

Aneurysm, genetic damage of

Aneurysm of sinus of Valsalva, genetic damage of

Angel shaped phalango epiphyseal dysplasia, genetic damage of

Angelman syndrome, genetic damage of

Angiofollicular ganglionic hyperplasia, genetic damage of

Angiofollicular lymph hyperplasia, genetic damage of

Angioimmunoblastic with dysproteinemia lymphadenopathy, genetic damage of

Angiokeratoma mental retardation coarse face, genetic damage of

Angioma hereditary neurocutaneous, genetic damage of

Angioneurotic edema hereditary due to C1 esterase deficiency, genetic damage of

Angiosarcoma of the liver, genetic damage of

Angiosarcoma of the scalp, genetic damage of

Angiotensin renin aldosterone hypertension, genetic damage of

Aniridia absent patella, genetic damage of

Aniridia ataxia renal agenesis psychomotor retardation, genetic damage of

Aniridia cerebellar ataxia mental deficiency, genetic damage of

Aniridia mental retardation syndrome, genetic damage of

Aniridia ptosis mental retardation obesity familial, genetic damage of

Aniridia renal agenesis psychomotor retardation, genetic damage of

Aniridia type 2, genetic damage of

Aniridia, sporadic, genetic damage of

Ankle defects short stature, genetic damage of

Ankyloblepharon cleft palate ectodermal defects, genetic damage of

Ankyloblepharon ectodermal defects cleft lip palate, genetic damage of

Ankyloblepharon filiforme adnatum cleft palate, genetic damage of

Ankyloblepharon filiforme imperforate anus, genetic damage of

Ankyloglossia heterochromia clasped thumbs, genetic damage of

Ankylosing spondylarthritis, genetic damage of

Ankylosing spondylitis, genetic damage of

Ankylosing vertebral hyperostosis with tylosis, genetic damage of

Ankylosis of teeth, genetic damage of

Annular constricting bands, genetic damage of

Annular pancreas, genetic damage of

Annuloaortic ectasia, genetic damage of

Ano-rectal atresia, genetic damage of

Anodontia, genetic damage of

Anonychia ectrodactyly, genetic damage of

Anonychia microcephaly, genetic damage of

Anonychia onychodystrophy, genetic damage of

Anonychia onychodystrophy brachydactyly type B, genetic damage of

Anophthalia pulmonary hypoplasia, genetic damage of

Anophthalmia cleft lip palate hypothalamic disorder, genetic damage of

Anophthalmia cleft palate micrognathia, genetic damage of

Anophthalmia esophageal atresia cryptorchidism, genetic damage of

Anophthalmia megalocornea cardiopathy skeletal anomalies, genetic damage of

Anophthalmia microcephaly hypogonadism, genetic damage of

Anophthalmia plus syndrome, genetic damage of

Anophthalmia short stature obesity, genetic damage of

Anophthalmia Waardenburg syndrome, genetic damage of

Anophthalmos with limb anomalies, genetic damage of

Anophthalmos, clinical, genetic damage of

Anorchia, genetic damage of

Anorchidism, genetic damage of

Anorectal anomalies, genetic damage of

Anorexia nervosa, genetic damage of

Anosmia, genetic damage of

Anotia, genetic damage of

Anotia facial palsy cardiac defect, genetic damage of

Ansell Bywaters Elderking syndrome, genetic damage of

Anterior horn disease, genetic damage of

Anterior pituitary insufficiency, familial, genetic damage of

Anthophobia, genetic damage of (possible)

Anti-factor VIII autoimmunization, genetic damage of

Anti-HLA hyperimmunization, genetic damage of

Anti-plasmin deficiency, congenital, genetic damage of

Antigen-peptide-transporter 2 deficiency, genetic damage of

Antinolo Nieto Borrego syndrome, genetic damage of

Antiphospholipid syndrome, genetic damage of

Antisocial personality disorder, genetic damage of

Antisynthetase syndrome, genetic damage of

Antithrombin deficiency, congenital, genetic damage of

Antley-Bixler syndrome, genetic damage of

Antlophobia, genetic damage of (possible)

Anyane Yeboa syndrome, genetic damage of

Aorta-pulmonary artery fistula, genetic damage of

Aortic aneurysm, genetic damage of

Aortic arch anomaly peculiar facies mental retardation, genetic damage of

Aortic arch interruption, genetic damage of

Aortic arches defect, genetic damage of

Aortic coarctation, genetic damage of

Aortic dissection lentiginosis, genetic damage of

Aortic supravalvular stenosis, genetic damage of

Aortic valve stenosis, genetic damage of

Aortic valves stenosis of the child, genetic damage of

Aortic window, genetic damage of

APECED syndrome, genetic damage of

Apert like polydactyly syndrome, genetic damage of

Apert syndrome, genetic damage of

Aphalangia hemivertebrae, genetic damage of

Aphalangia syndactyly microcephaly, genetic damage of

Aphthous stomatitis, genetic damage of

Apiphobia, genetic damage of (possible)

Aplasia cutis autosomal recessive, genetic damage of

Aplasia cutis cleft palate epidermolysis, genetic damage of

Aplasia cutis congenita, genetic damage of

Aplasia cutis congenita dominant, genetic damage of

Aplasia cutis congenita epibulbar dermoids, genetic damage of

Aplasia cutis congenita intestinal lymphangiectasia, genetic damage of

Aplasia cutis congenita of limbs recessive, genetic damage of

Aplasia cutis congenita recessive, genetic damage of

Aplasia cutis myopia, genetic damage of

Aplastic anemia, genetic damage of

Apo A-I deficiency, genetic damage of

Apolipoprotein C-II deficiency, genetic damage of

Apparent mineralocorticoid excess, genetic damage of

Apple peel syndrome, genetic damage of

Apraxia, buccofacial, genetic damage of

Apraxia, classic, genetic damage of

Apraxia, constructional, genetic damage of

Apraxia, ideational, genetic damage of

Apraxia, ideokinetic, genetic damage of

Apraxia, ideomotor, genetic damage of

Apraxia, motor, genetic damage of

Apraxia, oculomotor, genetic damage of

Apudoma, genetic damage of

Aqueductal stenosis, X linked, genetic damage of

Arachnodactyly ataxia cataract aminoaciduria mental retardation, genetic damage of

Arachnodactyly mental retardation dysmorphism, genetic damage of

Arachnoid cysts, genetic damage of

Arachnoiditis, genetic damage of

Arakawa'sa syndrome II, genetic damage of

Arc syndrome, genetic damage of

Aredyld syndrome, genetic damage of

Arginase deficiency, genetic damage of

Argininosuccinate synthetase deficiency, genetic damage of

Argininosuccinic aciduria, genetic damage of

Arhinia choanal atresia microphthalmia, genetic damage of

Arithmophobia, genetic damage of (possible)

Arnold Stckler Bourne syndrome, genetic damage of

Arnold-Chiari malformation, genetic damage of

Arnold-Chiari syndrome, genetic damage of

Aromatase deficiency, genetic damage of

Aromatic l amino acid decarboxylase deficiency, genetic damage of

Aromatic L-amino acid decarboxylase deficiency, genetic damage of

Arrhinia, genetic damage of

Arrhythmogenic right ventricular dysplasia, genetic damage of

Arroyo Garcia Cimadevilla syndrome, genetic damage of

Arrythmogenic right ventricular dysplasia, familial, genetic damage of

Arterial dysplasia, genetic damage of

Arterial tortuosity, genetic damage of

Arteriovenous malformation, genetic damage of

Arthritis, genetic damage of

Arthritis short stature deafness, genetic damage of

Arthritis, juvenile, genetic damage of

Arthrogryposis, genetic damage of

Arthrogryposis due to muscular dystrophy, genetic damage of

Arthrogryposis ectodermal dysplasia other anomalies, genetic damage of

Arthrogryposis epileptic seizures migrational brain disorder, genetic damage of

Arthrogryposis IUGR thoracic dystrophy, genetic damage of

Arthrogryposis like disorder, genetic damage of

Arthrogryposis like hand anomaly sensorineural, genetic damage of

Arthrogryposis multiplex congenita, genetic damage of

Arthrogryposis multiplex congenita CNS calcification, genetic damage of

Arthrogryposis multiplex congenita distal, genetic damage of

Arthrogryposis multiplex congenita distal type 1, genetic damage of

Arthrogryposis multiplex congenita distal type 2, genetic damage of

Arthrogryposis multiplex congenita neurogenic type, genetic damage of

Arthrogryposis multiplex congenita pulmonary hypoplasia, genetic damage of

Arthrogryposis multiplex congenita whistling face, genetic damage of

Arthrogryposis ophthalmoplegia retinopathy, genetic damage of

Arthrogryposis renal dysfunction cholestasis syndrome, genetic damage of

Arthrogryposis spinal muscular atrophy, genetic damage of

Artrogriposis, genetic damage of

Ascher's syndrome, genetic damage of

Asherman's syndrome, genetic damage of

Aspartylglycosaminuria, genetic damage of

Asperger syndrome, genetic damage of

Asphyxia neonatorum, genetic damage of

Asphyxiating thoracic dystrophy, genetic damage of

Asthenia, genetic damage of

Asthenophobia, genetic damage of (possible)

Astrocytoma, genetic damage of

Asymmetric septal hypertrophy, genetic damage of

Ataxia telangiectasia, genetic damage of

Ataxia telangiectasia variant V1, genetic damage of

Ataxia, Marie's, genetic damage of

Ataxiophobia, genetic damage of (possible)

Ataxophobia, genetic damage of (possible)

Athetosis, genetic damage of

Atkin Flaitz Patil Smith syndrome, genetic damage of

ATR-X, genetic damage of

Atresia of small intestine, genetic damage of

Atrial myxoma, familial, genetic damage of

Atrial septal defects, genetic damage of

Atrioventricular septal defect, genetic damage of

Atrophoderma of Pierini and Pasini, genetic damage of

Attention Deficit Disorder with Hyperactivity, genetic damage of

Atychiphobia, genetic damage of (possible)

Atypical lipodystrophy, genetic damage of

Auditory Perceptual disorder, genetic damage of

Aughton syndrome, genetic damage of

Ausems Wittebol Post Hennekam syndrome, genetic damage of

Autism, genetic damage of

Autoimmune hemolytic anemia, genetic damage of

Autoimmune hepatitis, genetic damage of

Autoimmune peripheral neuropathy, genetic damage of

Automysophobia, genetic damage of (possible)

Autonomic dysfunction, genetic damage of

Autonomic nervous system diseases, genetic damage of

Axial mesodermal dysplasia spectrum, genetic damage of

Axial osteosclerosis, genetic damage of

Ayazi syndrome, genetic damage of

B

B-cell lymphomas, genetic damage of

Bacillophobia, genetic damage of (possible)

Bader syndrome, genetic damage of

Baelz syndrome, genetic damage of

Bagatelle Cassidy syndrome, genetic damage of

Bahemuka Brown syndrome, genetic damage of

Baker Vinters syndrome, genetic damage of

Baker-Winegard syndrome, genetic damage of

Ballard syndrome, genetic damage of

Baller-Gerold syndrome, genetic damage of

Ballinger-Wallace syndrome, genetic damage of

Ballistophobia, genetic damage of (possible)

Balo disease, genetic damage of

Balo's concentric sclerosis, genetic damage of

Bamforth syndrome, genetic damage of

BANF acoustic neurinoma, genetic damage of

Bangstad syndrome, genetic damage of

Banki syndrome, genetic damage of

Bannayan-Zonana syndrome, genetic damage of

Bantil's syndrome, genetic damage of

Baraitser Brett Piesowicz syndrome, genetic damage of

Baraitser Burn Fixen syndrome, genetic damage of

Baraitser Rodeck Garner syndrome, genetic damage of

Barber Say syndrome, genetic damage of

Bardet-Biedl syndrome, type 1, genetic damage of

Bardet-Biedl syndrome, type 2, genetic damage of

Bardet-Biedl syndrome, type 3, genetic damage of

Bardet-Biedl syndrome, type 4, genetic damage of

Bare lymphocyte syndrome, genetic damage of

Barnicoat Baraitser syndrome, genetic damage of

Barophobia, genetic damage of (possible)

Barrett esophagus, genetic damage of

Barrow Fitzsimmons syndrome, genetic damage of

Bart Pumphrey syndrome, genetic damage of

Barth syndrome, genetic damage of

Bartsocas Papa syndrome, genetic damage of

Bartter syndrome, genetic damage of

Bartter syndrome, antenatal form, genetic damge of

Bartter's disease, genetic damage of

Basal cell nevus anodontia abnormal bone mineralization, genetic damage of

Basal ganglia diseases, genetic damage of

Basan syndrome, genetic damage of

Basaran Yilmaz syndrome, genetic damage of

Basedow's coma, genetic damage of

Basilar artery migraines, genetic damage of

Basilar impression primary, genetic damage of

Bassoe syndrome, genetic damage of

Bathophobia, genetic damage of (possible)

Batrachophobia, genetic damage of (possible)

Battaglia Neri syndrome, genetic damage of

Batten disease, genetic damage of

Batten Turner muscular dystrophy, genetic damage of

Baughman syndrome, genetic damage of

Bazex-Dupre-Christol syndrome, genetic damage of

Bazopoulou Kyrkanidou syndrome, genetic damage of

Bd syndrome, genetic damage of

Beals syndrome, genetic damage of

Beardwell syndrome, genetic damage of

Bébé Collodion syndrome, genetic damage of

Becker disease, genetic damage of

Becker's muscular dystrophy, genetic damage of

Becker's nevus, genetic damage of

Beckwith-Wiedemann syndrome, genetic damage of

Beemer Ertbruggen syndrome, genetic damage of

Beemer Langer syndrome, genetic damage of

Behcet syndrome, genetic damage of

Behr syndrome, genetic damage of

Behrens Baumann Dust syndrome, genetic damage of

Bell's palsy, genetic damage of

Bellini Chiumello Rinoldi syndrome, genetic damage of

Ben Ari Shuper Mimouni syndrome, genetic damage of

Benallegue Lacete syndrome, genetic damage of

Bencze syndrome, genetic damage of

Benign astrocytoma, genetic damage of

Benign autosomal dominant myopathy, genetic damage of

Benign congenital hypotonia, genetic damage of

Benign essential blepharospasm, genetic damage of

Benign essential tremor syndrome, genetic damage of

Benign familial hematuria, genetic damage of

Benign familial infantile convulsions, genetic damage of

Benign familial infantile epilepsy, genetic damage of

Benign familial pemphigus, genetic damage of

Benign lymphoma, genetic damage of

Benign mucosal pemphigoid, genetic damage of

Benign paroxysmal positional vertigo (BPPV), genetic damage of

Bennion Patterson syndrome, genetic damage of

Bentham Driessen Hanveld syndrome, genetic damage of

Beradinelli syndrome, genetic damage of

Berardinelli-Seip congenital lipodystrophy, genetic damage of

Berdon syndrome, genetic damage of

Berger disease, genetic damage of

Berk Tabatznik syndrome, genetic damage of

Berlin Breakage syndrome, genetic damage of

Bernard-Soulier syndrome, genetic damage of

Besnier-Boeck-Schaumann disease, genetic damage of

Beta-galactosidase-1 deficiency, genetic damage of

Beta-mannosidosis, genetic damage of

Beta-sarcoglycanopathy, genetic damage of

Beta-thalassemia, genetic damage of

Beta-thalassemia major anemia, genetic damage of

Betaketothiolase deficiency, genetic damage of

Bethlem myopathy, genetic damage of

Bhaskar Jagannathan syndrome, genetic damage of

Bibliophobia, genetic damage of (possible)

Bickel Fanconi glycogenosis, genetic damage of

Bicuspid aortic valve, genetic damage of

Bidirectional tachycardia, genetic damage of

BIDS syndrome, genetic damage of

Biemond syndrome, genetic damage of

Biemond syndrome type 1, genetic damage of

Biemond syndrome type 2, genetic damage of

Biermer disease, genetic damage of

Bifid nose dominant, genetic damage of

Bilateral acoustic neurofibromatosis, genetic damage of

Bilateral renal agenesis (Potter syndrome), genetic damage of

Bilateral renal agenesis dominant type, genetic damage of

Biliary atresia (generic term), genetic damage of

Biliary atresia, extrahepatic, genetic damage of

Biliary atresia, intrahepatic, non syndromic form, genetic damage of

Biliary atresia, intrahepatic, syndromic form, genetic damage of

Biliary cirrhosis, genetic damage of

Biliary malformation renal tubular insufficiency, genetic damage of

Biliary tract cancer, genetic damage of

Billard Toutain Maheut syndrome, genetic damage of

Billet Bear syndrome, genetic damage of

Bindewald Ulmer Muller syndrome, genetic damage of

Binswanger's disease, genetic damage of

Biotinidase deficiency, genetic damage of

Bird headed dwarfism Montreal type, genetic damage of

Birdshot chorioretinopathy, genetic damage of

Bixler Christian Gorlin syndrome, genetic damage of

Bjornstad syndrome, genetic damage of

Blackfan-Diamond anemia, genetic damage of

Bladder cancer, genetic damage of

Bladder neoplasm, genetic damage of

Blaichman syndrome, genetic damage of

Blastoma, genetic damage of

Blepharo cheilo dontic syndrome, genetic damage of

Blepharo facio skeletal syndrome, genetic damage of

Blepharo naso facial syndrome Van maldergem type, genetic damage of

Blepharonasofacial malformation syndrome, genetic damage of

Blepharophimosis, genetic damage of

Blepharophimosis nasal groove growth retardation, genetic damage of

Blepharophimosis ptosis esotropia syndactyly short, genetic damage of

Blepharophimosis ptosis syndactyly mental retardation, genetic damage of

Blepharophimosis radioulnar synostosis, genetic damage of

Blepharophimosis syndrome Ohdo type, genetic damage of

Blepharophimosis, ptosis, epicanthus inversus, genetic damage of

Blepharoptosis aortic anomaly, genetic damage of

Blepharoptosis cleft palate ectrodactyly dental anomalies, genetic damage of

Blepharoptosis myopia ectopia lentis, genetic damage of

Blepharospasm, genetic damage of

Blethen Wenick Hawkins syndrome, genetic damage of

Blomstrand syndrome, genetic damage of

Blood platelet disorders, genetic damage of

Blood vessel disorder, genetic damage of

Bloom syndrome, genetic damage of

Blount disease, genetic damage of

Blue cone monochromatism, genetic damage of

Blue diaper syndrome, genetic damage of

Blue rubber bleb nevus, genetic damage of

Bod syndrome, genetic damage of

Bone development disorder, genetic damage of

Bone dysplasia Azouz type, genetic damage of

Bone dysplasia corpus callosum agenesis, genetic damage of

Bone dysplasia lethal Holmgren type, genetic damage of

Bone dysplasia Moore type, genetic damage of

Bone fragility craniosynostosis proptosis hydrocephalus, genetic damage of

Bone marrow failure, genetic damage of

Bone marrow failure neurologic abnormalities, genetic damage of

Bone neoplasms, genetic damage of

Bone tumor (generic term), genetic damage of

Bonneau-Beaumont syndrome, genetic damage of

Bonneman Meinecke Reich syndrome, genetic damage of

Bonnemann Meinecke syndrome, genetic damage of

Bonnevie Ullrich Turner syndrome, genetic damage of

Book syndrome, genetic damage of

Boomerang dysplasia, genetic damage of

Booth Haworth Dilling syndrome, genetic damage of

BOR syndrome, genetic damage of

Borjeson syndrome, genetic damage of

Borjeson-Forssman-Lehmann syndrome, genetic damage of

Bork Stender Schmidt syndrome, genetic damage of

Borrone Di Rocco Crovato syndrome, genetic damage of

Boscherini Galasso Manca Bitti syndrome, genetic damage of

Bosma Henkin Christiansen syndrome, genetic damage of

Boucher Neuhauser syndrome, genetic damage of

Boudhina Yedes Khiari syndrome, genetic damage of

Bourneville syndrome, genetic damage of

Bourneville syndrome, type 1, genetic damage of

Bourneville syndrome, type 2, genetic damage of

Bowen syndrome, genetic damage of

Bowen's disease, genetic damage of

Bowen-Conradi syndrome, genetic damage of

Bowing congenital short bones, genetic damage of

Bowing of long bones congenital, genetic damage of

Boylan Dew Greco syndrome, genetic damage of

BPPV (benign paroxysmal positional vertigo), genetic damage of

Brachioskeletogenital syndrome, genetic damage of

Brachman-de Lange syndrome, genetic damage of

Brachycephalofrontonasal dysplasia, genetic damage of

Brachycephaly deafness cataract mental retardation, genetic damage of

Brachydactylie types b and e combined, genetic damage of

Brachydactylous dwarfism Mseleni type, genetic damage of

Brachydactyly absence of distal phalanges, genetic damage of

Brachydactyly anonychia, genetic damage of

Brachydactyly clinodactyly, genetic damage of

Brachydactyly dwarfism mental retardation, genetic damage of

Brachydactyly elbow wrist dysplasia, genetic damage of

Brachydactyly hypertension, genetic damage of

Brachydactyly long thumb type, genetic damage of

Brachydactyly mesomelia mental retardation heart defects, genetic damage of

Brachydactyly Mohr Wriedt type, genetic damage of

Brachydactyly nystagmus cerebellar ataxia, genetic damage of

Brachydactyly preaxial hallux varus, genetic damage of

Brachydactyly scoliosis carpal fusion, genetic damage of

Brachydactyly small stature face anomalies, genetic damage of

Brachydactyly Smorgasbord type, genetic damage of

Brachydactyly Temtamy type, genetic damage of

Brachydactyly tibial hypoplasia, genetic damage of

Brachydactyly type a1, genetic damage of

Brachydactyly type a2, genetic damage of

Brachydactyly type a3, genetic damage of

Brachydactyly type a4, genetic damage of

Brachydactyly type a5 nail dysplasia, genetic damage of

Brachydactyly type a6, genetic damage of

Brachydactyly type a7, genetic damage of

Brachydactyly type b, genetic damage of

Brachydactyly type c, genetic damage of

Brachydactyly type e, genetic damage of

Brachymesomelia renal syndrome, genetic damage of

Brachymesophalangy 2 and 5, genetic damage of

Brachymesophalangy mesomelic short limbs osseous anomalies, genetic damage of

Brachymesophalangy type 2, genetic damage of

Brachymetapody anodontia hypotrichosis albinoidism, genetic damage of

Brachymorphism onychodysplasia dysphalangism syndrome, genetic damage of

Brachyolmia, genetic damage of

Brachyolmia recessive Hobaek type, genetic damage of

Brachytelephalangy characteristic facies Kallmann, genetic damage of

Braddock Carey syndrome, genetic damage of

Braddock Jones Superneau syndrome, genetic damage of

Bradykinesia, genetic damage of

Brain cavernous angioma, genetic damage of

Brain neoplasms, genetic damage of

Brain stem neoplasms, genetic damage of

Branched chain ketoaciduria, genetic damage of

Branchial arch defects, genetic damage of

Branchial arch syndrome X linked, genetic damage of

Branchial dysplasia mental retardation inguinal hernia, genetic damage of

Branchio oculo facial syndrome Hing type, genetic damage of

Branchio-oculo-facial syndrome, genetic damage of

Branchiootorenal syndrome, genetic damage of

Breast and ovarian cancer, genetic damage of

Breast cancer, familial, genetic damage of

Brittle bone disease, genetic damage of

Brittle bone syndrome lethal type, genetic damage of

Brittle cornea syndrome, genetic damage of

Broad beta disease, genetic damage of

Broad-betalipoproteinemia, genetic damage of

Bromidrosiphobia, genetic damage of (possible)

Bronchiectasis oligospermia, genetic damage of

Bronchiolitis obliterans with obstructive pulmonary disease, genetic damage of

Bronchiolotis obliterans organizing pneumonia (BOOP), genetic damage of

Bronchogenic cyst, genetic damage of

Bronchopulmonary amyloidosis, genetic damage of

Bronchopulmonary dysplasia, genetic damage of

Brown syndrome, genetic damage of

Brown-Sequard syndrome, genetic damage of

Bruce Winship syndrome, genetic damage of

Brucellosis, genetic damage of

Bruck syndrome, genetic damage of

Brugada syndrome, genetic damage of

Brunner Winter syndrome, genetic damage of

Brunoni syndrome, genetic damage of

Bruton type agammaglobulinemia, genetic damage of

Bruyn Scheltens syndrome, genetic damage of

Budd-Chiari syndrome, genetic damage of

Buerger's disease, genetic damage of

Bufonophobia, genetic damage of (possible)

Bulbospinal amyotrophy, X linked, genetic damage of

Bulimia nervosa, genetic damage of

Bull Nixon syndrome, genetic damage of

Bullous dystrophy macular type, genetic damage of

Bullous ichtyosiform erythroderma congenita, genetic damage of

Bullous pemphigoid, genetic damage of

Buntinx Lormans Martin syndrome, genetic damage of

Burkitt's lymphoma, genetic damage of

Burn Goodship syndrome, genetic damage of

Burnett Schwartz Berberian syndrome, genetic damage of

Burning mouth syndrome, genetic damage of

Buschke Fischer Brauer syndrome, genetic damage of

Buschke Ollendorff syndrome, genetic damage of

Buschke's scleredema, genetic damage of

Bustos Simosa Pinto Cisternas syndrome, genetic damage of

Buttiens Fryns syndrome, genetic damage of

Butyrylcholinesterase deficiency, genetic damage of

C

C syndrome, genetic damage of

C1 esterase deficiency (type 2 with ascites), genetic damage of

Cacchi Ricci disease, genetic damage of

CACH syndrome, genetic damage of

Cacophobia, genetic damage of (possible)

CADASIL, genetic damage of

Cafe au lait spots syndrome, genetic damage of

Caffey disease, genetic damage of

Cahmr syndrome, genetic damage of

Calcinosis-Raynaud phenomenon-sclerodactyly-telangiectasia, genetic damage of

Calciphylaxis, genetic damage of

Calculi, genetic damage of (possible)

Calderon Gonzalez Cantu syndrome, genetic damage of

Calloso genital dysplasia, genetic damage of

Callus disease, genetic damage of

Calpainopathy, genetic damage of

Calvarial hyperostosis, genetic damage of

Camera Lituania Cohen syndrome, genetic damage of

Camfak syndrome, genetic damage of

Campomelia Cumming type, genetic damage of

Camptobrachydactyly, genetic damage of

Camptocormism, genetic damage of

Camptodactyly fibrous tissue hyperplasia skeletal dysplasia, genetic damage of

Camptodactyly joint contractures facial skeletal dysplasia, genetic damage of

Camptodactyly overgrowth unusual facies, genetic damage of

Camptodactyly syndrome Guadalajara type 1, genetic damage of

Camptodactyly syndrome Guadalajara type 2, genetic damage of

Camptodactyly taurinuria, genetic damage of

Camptodactyly vertebral fusion, genetic damage of

Camptomelic syndrome, genetic damage of

Camurati Engelmann disease, genetic damage of

Canavan leukodystrophy, genetic damage of

Cancrum oris, genetic damage of

Cantalamessa Baldini Ambrosi syndrome, genetic damage of

Cantu Sanchez Corona Fragoso syndrome, genetic damage of

Cantu Sanchez Corona Garcia syndrome, genetic damage of

Cantu Sanchez Corona Hernandes syndrome, genetic damage of

Capillary leak syndrome with monoclonal gammopathy, genetic damage of

Capillary venous leptomeningeal angiomatosis, genetic damage of

Capos syndrome, genetic damage of

Caratolo Cilio Pessagno syndrome, genetic damage of

Carbamoyl phosphate synthetase deficiency, genetic damage of

Carbamoyl phosphate synthetase deficiency (genetic form), genetic damage of

Carbamoyl-phosphate synthase I deficiency disease (ornithine carbamoyl phosphate deficiency), genetic damage of

Carbamyl phosphate synthetase deficiency, genetic damage of

Carbohydrate deficient glycoprotein syndrome, genetic damage of

Carbon baby syndrome, genetic damage of

Carbonic anhydrase II deficiency, genetic damage of

Carboxylase deficiency, multiple, genetic damage of

Carcinoid syndrome, genetic damage of

Carcinoma, genetic damage of

Carcinoma of the vocal tract, genetic damage of

Carcinoma, squamous cell, genetic damage of

Carcinophobia, genetic damage of (possible)

Cardiac and laterality defects, genetic damage of

Cardiac conduction defect, familial, genetic damage of

Cardiac diverticulum, genetic damage of

Cardiac hydatid cysts with intracavitary expansion, genetic damage of

Cardiac malformation, genetic damage of

Cardiac valvular dysplasia, X linked, genetic damage of

Cardio-facio-cutaneous syndrome, genetic damage of

Cardioauditory syndrome, genetic damage of

Cardiofacial syndrome short limbs, genetic damage of

Cardiogenital syndrome, genetic damage of

Cardiomelic syndrome Stratton Koehler type, genetic damage of

Cardiomyopathic lentiginosis, genetic damage of (possible)

Cardiomyopathy cataract hip spine disease, genetic damage of

Cardiomyopathy diabetes deafness, genetic damage of

Cardiomyopathy dilated with conduction defect type 1, genetic damage of

Cardiomyopathy dilated with conduction defect type 2, genetic damage of

Cardiomyopathy due to anthracyclines, genetic damage of

Cardiomyopathy hearing loss type t RNA lysine gene mutation, genetic damage of

Cardiomyopathy hypogonadism metabolic anomalies, genetic damage of

Cardiomyopathy spherocytosis, genetic damage of

Cardiomyopathy, familial dilated, genetic damage of

Cardiomyopathy, familial hypertrophic, genetic damage of

Cardiomyopathy, X linked, fatal infantile, genetic damage of

Cardiophobia, genetic damage of (possible)

Cardioskeletal myopathy-neutropenia, genetic damage of

Cardiospasm, genetic damage of

Carey Fineman Ziter syndrome, genetic damage of

Carnevale Canun Mendoza syndrome, genetic damage of

Carnevale Hernandez Castillo syndrome, genetic damage of

Carnevale Krajewska Fischetto syndrome, genetic damage of

Carney syndrome, genetic damage of

Carnitine palmitoyl transferase 1 deficiency, genetic damage of

Carnitine palmitoyl transferase 2 deficiency, genetic damage of

Carnitine palmityl transferase deficiency, genetic damage of

Carnitine transporter deficiency, genetic damage of

Carnitine-acylcarnitine translocase deficiency, genetic damage of

Carnophobia, genetic damage of (possible)

Carnosinase deficiency, genetic damage of

Carnosinemia, genetic damage of

Caroli disease, genetic damage of

Carpal deformity migrognathia microstomia, genetic damage of

Carpal tunnel syndrome, genetic damage of

Carpenter Hunter type, genetic damage of

Carpenter syndrome, genetic damage of

Carpo tarsal osteochondromatosis, genetic damage of

Carpo tarsal osteolysis recessive, genetic damage of

Carrington syndrome, genetic damage of

Cartilage hair hypoplasia like syndrome, genetic damage of

Cartilaginous neoplasms, genetic damage of

Cartwright Nelson Fryns syndrome, genetic damage of

Cassia Stocco Dos Santos syndrome, genetic damage of

Castleman's disease, genetic damage of

Castro Gago Pombo Novo syndrome, genetic damage of

Cat cry syndrome, genetic damage of

Cat eye syndrome, genetic damage of

Cat Rodrigues syndrome, genetic damage of

Cat scratch disease, genetic damage of

Catagelophobia, genetic damage of (possible)

Catapedaphobia, genetic damage of (possible)

Cataract aberrant oral frenula growth retardation, genetic damage of

Cataract alopecia sclerodactyly, genetic damage of

Cataract anterior polar dominant, genetic damage of

Cataract ataxia deafness, genetic damage of

Cataract cardiomyopathy, genetic damage of

Cataract congenital autosomal dominant, genetic damage of

Cataract congenital dominant non nuclear, genetic damage of

Cataract congenital ichthyosis, genetic damage of

Cataract congenital Volkmann type, genetic damage of

Cataract congenital with microphthalmia, genetic damage of

Cataract dental syndrome, genetic damage of

Cataract Hutterite type, genetic damage of

Cataract hypertrichosis mental retardation, genetic damage of

Cataract mental retardation hypogonadism, genetic damage of

Cataract microcornea syndrome, genetic damage of

Cataract microcornea X linked, genetic damage of

Cataract microphthalmia septal defect, genetic damage of

Cataract skeletal anomalies, genetic damage of

Cataract, total congenital, genetic damage of

Cataract-glaucoma, genetic damage of

CATCH 22 syndrome, genetic damage of

Catecholamine hypertension, genetic damage of

Catel Manzke syndrome, genetic damage of

Caudal appendage deafness, genetic damage of

Caudal duplication, genetic damage of

Caudal regression syndrome, genetic damage of

Causalgia, genetic damage of

Cavernous hemangioma, genetic damage of

Cavernous lymphangioma, genetic damage of

Cayler syndrome, genetic damage of

CBPS or Perisylvian syndrome, genetic damage of

CCA syndrome, genetic damage of

Ccge syndrome, genetic damage of

CDG syndrome, genetic damage of

CDG syndrome type 1A, genetic damage of

CDG syndrome type 1B, genetic damage of

CDG syndrome type 1C, genetic damage of

CDG syndrome type 2, genetic damage of

CDG syndrome type 3, genetic damage of

CDG syndrome type 4, genetic damage of

CDK4 linked melanoma, genetic damage of

Cecato De Lima Pinheiro syndrome, genetic damage of

Celiac disease epilepsy occipital calcifications, genetic damage of

Celiac sprue, genetic damage of

Cenani Lenz syndactylism, genetic damage of

Cennamo Gangemi syndrome, genetic damage of

Central core disease, genetic damage of

Central diabetes insipidus, genetic damage of

Central serous chorioretinopathy, genetic damage of

Central type neurofibromatosis, genetic damage of

Centromeric instability immunodeficiency syndrome, genetic damage of

Centronuclear myopathy, congenital, genetic damage of

Centrotemporal epilepsy, genetic damage of

Cephalopolysyndactyly, genetic damage of

Ceramidase deficiency, genetic damage of

Ceramide trihexosidosis, genetic damage of

Ceraunophobia, genetic damage of (possible)

Cerebellar agenesis, genetic damage of

Cerebellar ataxia, genetic damage of

Cerebellar ataxia areflexia pes cavus optic atrophy, genetic damage of

Cerebellar ataxia ectodermal dysplasia, genetic damage of

Cerebellar ataxia infantile with progressive external opht halmoplegia, genetic damage of

Cerebellar ataxia, dominant pure, genetic damage of

Cerebellar degeneration, genetic damage of

Cerebellar degeneration, subacute, genetic damage of

Cerebellar hypoplasia, genetic damage of

Cerebellar hypoplasia endosteal sclerosis, genetic damage of

Cerebellar hypoplasia tapetoretinal degeneration, genetic damage of

Cerebellar parenchymal degeneration, genetic damage of

Cerebellar vermis agenesis, genetic damage of

Cerebelloolivary atrophy, genetic damage of

Cerebelloparenchymal disorder 3, genetic damage of

Cerebellum agenesis hydrocephaly, genetic damage of

Cerebral amyloid angiopathy, genetic damage of

Cerebral aneurysm, genetic damage of

Cerebral calcification cerebellar hypoplasia, genetic damage of

Cerebral calcifications opalescent teeth phosphaturia, genetic damage of

Cerebral cavernous malformation, genetic damage of

Cerebral cavernous malformations, genetic damage of

Cerebral gigantism, genetic damage of

Cerebral gigantism jaw cysts, genetic damage of

Cerebral malformations hypertrichosis claw hands, genetic damage of

Cerebral palsy, genetic damage of

Cerebral ventricle neoplasm, genetic damage of

Cerebro facio articular syndrome, genetic damage of

Cerebro facio thoracic dysplasia, genetic damage of

Cerebro oculo dento auriculo skeletal syndrome, genetic damage of

Cerebro oculo facio skeletal syndrome, genetic damage of

Cerebro oculo genital syndrome, genetic damage of

Cerebro oculo skeleto renal syndrome, genetic damage of

Cerebro reno digital syndrome, genetic damage of

Cerebro-costo-mandibular syndrome, genetic damage of

Cerebro-oculo-facio-skeletal syndrome, genetic damage of

Cerebroarthrodigital syndrome, genetic damage of

Cerebroretinal vasculopathy, genetic damage of

Cerebrotendinous xanthomatosis (CTX), genetic damage of

Ceroid lipofuscinose, neuronal, genetic damage of

Ceroid lipofuscinose, neuronal 1, infantile, genetic damage of

Ceroid lipofuscinose, neuronal 2, late infantile, genetic damage of

Ceroid lipofuscinose, neuronal 3, juvenile, genetic damage of

Ceroid lipofuscinose, neuronal 4, adult type, genetic damage of

Ceroid lipofuscinose, neuronal 5, late infantile, genetic damage of

Ceroid lipofuscinose, neuronal 6, late infantile, genetic damage of

Cervical hypertrichosis neuropathy, genetic damage of

Cervical hypertrichosis peripheral neuropathy, genetic damage of

Cervical ribs sprengel anomaly polydactyly, genetic damage of

Cervical vertebral fusion, genetic damage of

Cervicooculoacoustic syndrome, genetic damage of

CFC syndrome, genetic damage of

Chagas disease, genetic damage of

Chalazion, genetic damage of

Chanarin disease, genetic damage of

Chanarin Dorfman syndrome ichthyosis, genetic damage of

Chandler's syndrome, genetic damage of

Chands syndrome, genetic damage of

Chang Davidson Carlson syndrome, genetic damage of

Chaotic atrial tachycardia, genetic damage of

Char syndrome, genetic damage of

Charcot disease, genetic damage of

Charcot Marie tooth disease deafness dominant type, genetic damage of

Charcot Marie tooth disease deafness mental retardation, genetic damage of

Charcot Marie tooth disease deafness recessive type, genetic damage of

Charcot Marie tooth type 1 aplasia cutis congenita, genetic damage of

Charcot-Marie-tooth disease, genetic damage of

Charcot-Marie-tooth disease, X linked type 2, recessive, genetic damage of

Charcot-Marie-tooth disease, X linked type 3, recessive, genetic damage of

Charcot-Marie-tooth disease type 1A, genetic damage of

Charcot-Marie-tooth disease type 1B, genetic damage of

Charcot-Marie-tooth disease type 1C, genetic damage of

Charcot-Marie-tooth disease type 2A, genetic damage of

Charcot-Marie-tooth disease type 2B, genetic damage of

Charcot-Marie-tooth disease type 2C, genetic damage of

Charcot-Marie-tooth disease type 2D, genetic damage of

Charcot-Marie-tooth disease type 4A, genetic damage of

Charcot-Marie-tooth disease, intermediate form, genetic damage of

Charcot-Marie-tooth disease, neuronal, type A, genetic damage of

Charcot-Marie-tooth disease, neuronal, type B, genetic damage of

Charcot-Marie-tooth disease, neuronal, type C, genetic damage of

Charcot-Marie-tooth disease, neuronal, type D, genetic damage of

Charcot-Marie-tooth peroneal muscular atrophy, X linked, genetic damage of

CHARGE association, genetic damage of

Charlie M syndrome, genetic damage of

Chavany-Brunhes syndrome, genetic damage of

Chediak-Higashi syndrome, genetic damage of

Chediak-Higashi-like syndrome, genetic damage of

Chemke Oliver Mallek syndrome, genetic damage of

Chemodectoma, genetic damage of

Chemophobia, genetic damage of (possible)

Chen Kung Ho Kaufman Mcalister syndrome, genetic damage of

Cherubism, genetic damage of

Cherubism gingival fibromatosis mental retardation, genetic damage of

Cherubism optic atrophy short stature, genetic damage of

Chiari type 1 malformation, genetic damage of

Chiari-Frommel syndrome, genetic damage of

CHILD syndrome ichthyosis, genetic damage of

Childhood ataxia with diffuse central nervous system, genetic damage of

Childhood disintegrative disorder, genetic damage of

Childhood pustular psoriasis, genetic damage of

Chionophobia, genetic damage of (possible)

Chiraptophobia, genetic damage of (possible)

Chirophobia, genetic damage of (possible)

Chitayat Haj Chahine syndrome, genetic damage of

Chitayat Meunier Hodgkinson syndrome, genetic damage of

Chitayat Moore Del Bigio syndrome, genetic damage of

Chitty Hall Baraitser syndrome, genetic damage of

Chitty Hall Webb syndrome, genetic damage of

Choanal atresia deafness cardiac defects dysmorphia, genetic damage of

Cholangiocarcinoma, genetic damage of

Cholangitis, primary sclerosing, genetic damage of

Cholecystitis, genetic damage of

Choledochal cyst hand malformation, genetic damage of

Cholemia, familial, genetic damage of

Cholerophobia, genetic damage of (possible)

Cholestasis, genetic damage of

Cholestasis pigmentary retinopathy cleft palate, genetic damage of

Cholestasis, progressive familial intrahepatic, genetic damage of

Cholestasis, progressive familial intrahepatic 1, genetic damage of

Cholestasis, progressive familial intrahepatic 2, genetic damage of

Cholestasis, progressive familial intrahepatic 3, genetic damage of

Cholestatic jaundice renal tubular insufficiency, genetic damage of

Cholesterol ester storage disease, genetic damage of

Cholesterol esterification disorder, genetic damage of

Chondroblastoma (benign), genetic damage of

Chondrocalcinosis, genetic damage of

Chondrocalcinosis familial articular, genetic damage of

Chondrodysplasia lethal recessive, genetic damage of

Chondrodysplasia pseudohermaphrodism syndrome, genetic damage of

Chondrodysplasia punctata, genetic damage of

Chondrodysplasia punctata with steroid sulfatase deficiency, genetic damage of

Chondrodysplasia punctata, brachytelephalangic, genetic damage of

Chondrodysplasia punctata, rhizomelic form, genetic damage of

Chondrodysplasia punctata, Sheffield type, genetic damage of

Chondrodysplasia situs inversus imperforate anus polydactyly, genetic damage of

Chondrodysplasia, Grebe type, genetic damage of

Chondrodystrophy, genetic damage of

Chondroectodermal dysplasia, genetic damage of

Chondroma (benign), genetic damage of

Chondromalacia, genetic damage of

Chondromatosis (benign), genetic damage of

Chondrosarcoma (malignant), genetic damage of

Chondrysplasia punctata, humero-metacarpal type, genetic damage of

Chordoma, genetic damage of

Chorea, genetic damage of

Chorea acanthocytosis, genetic damage of

Chorea familial benign, genetic damage of

Chorea minor, genetic damage of

Choreoacanthocytosis amyotrophic, genetic damage of

Choreoathetosis familial paroxysmal, genetic damage of

Choreoathetosis hyperuricemia, genetic damage of

Choriocarcinoma, genetic damage of

Chorioretinopathy dominant form microcephaly, genetic damage of

Choroid plexus cyst, genetic damage of

Choroid plexus neoplasms, genetic damage of

Choroidal atrophy alopecia, genetic damage of

Choroideremia, genetic damage of

Choroideremia hypopituitarism, genetic damage of

Choroiditis, genetic damage of

Choroiditis, serpiginous, genetic damage of

Choroido cerebral calcification syndrome infantile, genetic damage of

Chorophobia, genetic damage of (possible)

Christian Demyer Franken syndrome, genetic damage of

Christian Johnson Angenieta syndrome, genetic damage of

Christian syndrome, genetic damage of

Christianson Fourie syndrome, genetic damage of

Christmas disease, genetic damage of

Chrometophobia, genetic damage of (possible)

Chromophobe renal carcinoma, genetic damage of

Chromophobia, genetic damage of (possible)

Chromosomal triplication, genetic damage of

Chromosome 1 ring, genetic damage of

Chromosome 10 ring, genetic damage of

Chromosome 10, distal trisomy 10q, genetic damage of

Chromosome 10, monosomy 10p, genetic damage of

Chromosome 11, partial monosomy 11q, genetic damage of

Chromosome 11, partial trisomy 11q, genetic damage of

Chromosome 11-14 translocation, genetic damage of

Chromosome 11q syndrome, genetic damage of

Chromosome 12 ring, genetic damage of

Chromosome 12p deletion, genetic damage of

Chromosome 13 ring, genetic damage of

Chromosome 13, partial monosomy 13q, genetic damage of

Chromosome 13q syndrome, genetic damage of

Chromosome 13q-mosaicism, genetic damage of

Chromosome 14 ring, genetic damage of

Chromosome 14 trisomy, genetic damage of

Chromosome 14, trisomy mosaic, genetic damage of

Chromosome 15 ring, genetic damage of

Chromosome 15, distal trisomy 15q, genetic damage of

Chromosome 17deletion, genetic damage of

Chromosome 18 long arm deletion syndrome, genetic damage of

Chromosome 18, monosomy 18p, genetic damage of

Chromosome 18, ring, genetic damage of

Chromosome 18, tetrasomy 18p, genetic damage of

Chromosome 18p-syndrome, genetic damage of

Chromosome 18q-syndrome, genetic damage of

Chromosome 19 ring, genetic damage of

Chromosome 1p36 depletion syndrome, genetic damage of

Chromosome 20 ring, genetic damage of

Chromosome 21 ring, genetic damage of

Chromosome 22 ring, genetic damage of

Chromosome 22 trisomy, genetic damage of

Chromosome 22, trisomy mosaic, genetic damage of

Chromosome 3 deletion of distal, genetic damage of

Chromosome 3 duplication syndrome, genetic damage of

Chromosome 3, monosomy 3p, genetic damage of

Chromosome 3, monosomy 3p2, genetic damage of

Chromosome 3, Trisomy 3q2, genetic damage of

Chromosome 4 ring, genetic damage of

Chromosome 4 short arm deletion, genetic damage of

Chromosome 4 Trisomy, genetic damage of

Chromosome 4, monosomy 4q, genetic damage of

Chromosome 4, monosomy distal 4q, genetic damage of

Chromosome 4, Partial Trisomy Distal 4q, genetic damage of

Chromosome 4, Trisomy 4p, genetic damage of

Chromosome 4q-syndrome, genetic damage of

Chromosome 5 trisomy 5p, genetic damage of

Chromosome 5p-syndrome, genetic damage of

Chromosome 6 ring, genetic damage of

Chromosome 6, partial trisomy 6q, genetic damage of

Chromosome 7 ring, genetic damage of

Chromosome 7, monosomy 7p2, genetic damage of

Chromosome 8 deletion, genetic damage of

Chromosome 8 ring, genetic damage of

Chromosome 8, monosomy 8p2, genetic damage of

Chromosome 9 ring, genetic damage of

Chromosome 9, partial monosomy 9p, genetic damage of

Chromosome 9, tetrasomy 9p, genetic damage of

Chromosome 9, trisomy 9p (multiple variants), genetic damage of

Chromosome 9, trisomy mosaic, genetic damage of

Chromosome disorders, genetic damage of

Chromosome triploidy syndrome, genetic damage of

Chronic demyelinizing neuropathy with IgM monoclonal, genetic damage of

Chronic erosive gastritis, genetic damage of

Chronic fatigue immune dysfunction syndrome, genetic damage of

Chronic granulomatous disease, genetic damage of

Chronic hiccup, genetic damage of (possible)

Chronic inflammatory demyelinating polyneuropathy, genetic damage of

Chronic lymphocytic leukemia, genetic damage of

Chronic myelogenous leukemia, genetic damage of

Chronic myelomonocytic leukemia, genetic damage of

Chronic necrotizing vasculitis, genetic damage of

Chronic neutropenia, genetic damage of

Chronic polyradiculoneuritis, genetic damage of

Chronic recurrent multifocal osteomyelitis, genetic damage of

Chronic renal failure, genetic damage of

Chronic spasmodic dysphonia, genetic damage of

Chronic, infantile, neurological, cutaneous, articular syndrome, genetic damage of

Chronomentrophobia, genetic damage of (possible)

Chudley Lowry Hoar syndrome, genetic damage of

Chudley Rozdilsky syndrome, genetic damage of

Chudley-Mccullough syndrome, genetic damage of

Churg-Strauss syndrome, genetic damage of

Chylous ascites, genetic damage of

Cicatricial pemphigoid, genetic damage of

Ciliary discoordination, due to random ciliary orientation, genetic damage of

Ciliary dyskinesia, due to transposition of ciliary microtubules, genetic damage of

Ciliary dyskinesia-bronchiectasis, genetic damage of

Cilliers Beighton syndrome, genetic damage of

Cinca syndrome, genetic damage of

Circumscribed cutaneous aplasia of the vertex, genetic damage of

Circumscribed disseminated keratosis Jadassohn Lew type, genetic damage of

Citrullinemia, genetic damage of

Clarkson disease, genetic damage of

Clayton Smith Donnai syndrome, genetic damage of

Cleft hand absent tibia, genetic damage of

Cleft lip, genetic damage of

Cleft lip and palate malrotation cardiopathy, genetic damage of

Cleft lip and/or palate with mucous cysts of lower lip, genetic damage of

Cleft lip palate abnormal thumbs microcephaly, genetic damage of

Cleft lip palate deafness sacral lipoma, genetic damage of

Cleft lip palate dysmorphism Kumar type, genetic damage of

Cleft lip palate ectrodactyly, genetic damage of

Cleft lip palate facial eye heart intestinal anomalies, genetic damage of

Cleft lip palate incisor and finger anomalies, genetic damage of

Cleft lip palate lip pits limb deficiency, genetic damage of

Cleft lip palate mental retardation corneal opacity, genetic damage of

Cleft lip palate oligodontia syndactyly pili torti, genetic damage of

Cleft lip palate pituitary deficiency, genetic damage of

Cleft lip palate-tetraphocomelia, genetic damage of

Cleft lip with or without cleft palate, genetic damage of

Cleft lower lip cleft lateral canthi chorioretinal, genetic damage of

Cleft palate, genetic damage of

Cleft palate cardiac defect ectrodactyly, genetic damage of

Cleft palate colobomata radial synostosis deafness, genetic damage of

Cleft palate heart disease polydactyly absent tibia, genetic damage of

Cleft palate lateral synechia syndrome, genetic damage of

Cleft palate short stature vertebral anomalies, genetic damage of

Cleft palate stapes fixation oligodontia, genetic damage of

Cleft palate X linked, genetic damage of

Cleft tongue syndrome, genetic damage of

Cleft upper lip median cutaneous polyps, genetic damage of

Clefting ectropion conical teeth, genetic damage of

Cleido rhizomelic syndrome, genetic damage of

Cleidocranial dysplasia, genetic damage of

Cleidocranial dysplasia micrognathia absent thumbs, genetic damage of

Cleisiophobia, genetic damage of (possible)

Climacophobia, genetic damage of (possible)

Clinophobia, genetic damage of (possible)

Cloacal exstrophy, genetic damage of

Clouston syndrome, genetic damage of

Cloverleaf skull bone dysplasia, genetic damage of

Cloverleaf skull micromelia thoracic dysplasia, genetic damage of

Cloverleaf skull syndrome, genetic damage of

Cluster headache, genetic damage of

Coach syndrome, genetic damage of

Coarctation of aorta dominant, genetic damage of

Coarse face hypotonia constipation, genetic damage of

Coats disease, genetic damage of

Cochin Jewish disorder, genetic damage of

Cockayne syndrome type 1, genetic damage of

Cockayne syndrome type 2, genetic damage of

Cockayne syndrome type 3, genetic damage of

Cockayne's syndrome, genetic damage of

Codas syndrome, genetic damage of

Coenzyme Q cytochrome c reductase, deficiency of, genetic damage of

Coffin-Lowry syndrome, genetic damage of

Coffin-Siris syndrome, genetic damage of

Cofs syndrome, genetic damage of

Cogan's syndrome, genetic damage of

Cogan-Reese syndrome, genetic damage of

Cohen Hayden syndrome, genetic damage of

Cohen Lockood Wyborney syndrome, genetic damage of

Cohen syndrome, genetic damage of

Coimetrophobia, genetic damage of (possible)

Colangite esclerosante por paracoccidiodomicose, genetic damage of

Colavita Kozlowski syndrome, genetic damage of

Cold agglutination syndrome, genetic damage of

Cold agglutinin disease, genetic damage of

Cold antibody hemolytic anemia, genetic damage of

Cold contact urticaria, genetic damage of

Cold urticaria, genetic damage of

Cole carpenter syndrome, genetic damage of

Coleman Randall syndrome, genetic damage of

Colitis, genetic damage of (possible)

Collagen disorder, genetic damage of

Collagenous colitis, genetic damage of

Collins Pope syndrome, genetic damage of

Collins Sakati syndrome, genetic damage of

Coloboma chorioretinal cerebellar vermis aplasia, genetic damage of

Coloboma hair abnormality, genetic damage of

Coloboma of choroid and retina, genetic damage of

Coloboma of eye lens, genetic damage of

Coloboma of iris, genetic damage of

Coloboma of lens ala nasi, genetic damage of

Coloboma of macula, genetic damage of

Coloboma of macula type b brachydactyly, genetic damage of

Coloboma of optic papilla, genetic damage of

Coloboma porencephaly hydronephrosis, genetic damage of

Coloboma uveal with cleft lip palate and mental retardation, genetic damage of

Coloboma, ocular, genetic damage of

Colobomata unilobar lung heart defect, genetic damage of

Colobomatous microphthalmia, genetic damage of

Colobomatous microphthalmia heart disease hearing, genetic damage of

Colon cancer, familial nonpolyposis, genetic damage of

Colonic atresia, genetic damage of

Colver Steer Godman syndrome, genetic damage of

Combarros Calleja Leno syndrome, genetic damage of

Combined hyperlipidemia, familial, genetic damage of

Common mesentery, genetic damage of

Common variable immunodeficiency, genetic damage of

Compartment syndrome, genetic damage of

Complement component 2 deficiency, genetic damage of

Complement component receptor 1, genetic damage of

Complete atrioventricular canal, genetic damage of

Complex 1 mitochondrial respiratory chain deficiency, genetic damage of

Complex 2 mitochondrial respiratory chain deficiency, genetic damage of

Complex 3 mitochondrial respiratory chain deficiency, genetic damage of

Complex 4 mitochondrial respiratory chain deficiency, genetic damage of

Complex 5 mitochondrial respiratory chain deficiency, genetic damage of

Conductive deafness malformed external ear, genetic damage of

Conductive hearing loss, genetic damage of, genetic damage of

Cone dystrophy, genetic damage of

Cone rod dystrophy, genetic damage of

Cone rod dystrophy amelogenesis imperfecta, genetic damage of

Congenital absence of the uterus and vagina, genetic damage of

Congenital adrenal hyperplasia type 1, genetic damage of

Congenital adrenal hyperplasia type 2, genetic damage of

Congenital adrenal hyperplasia type 3, genetic damage of

Congenital adrenal hyperplasia type 4, genetic damage of

Congenital adrenal hyperplasia type 5, genetic damage of

Congenital afibrinogenemia, genetic damage of

Congenital alopecia X linked, genetic damage of

Congenital amputation, genetic damage of

Congenital aneurysms of the great vessels, genetic damage of

Congenital antithrombin III deficiency, genetic damage of

Congenital aplastic anemia, genetic damage of

Congenital arteriovenous shunt, genetic damage of

Congenital articular rigidity, genetic damage of

Congenital benign spinal muscular atrophy dominant, genetic damage of

Congenital brain disorder, genetic damage of

Congenital bronchobiliary fistula, genetic damage of

Congenital cardiovascular disorder, genetic damage of

Congenital cardiovascular malformations, genetic damage of

Congenital cardiovascular shunt, genetic damage of

Congenital constricting band, genetic damage of

Congenital contractual arachnodactyly, genetic damage of

Congenital contractures, genetic damage of

Congenital craniosynostosis maternal hyperthyroiditis, genetic damage of

Congenital cystic adenomatoid malformation, genetic damage of

Congenital cystic eye multiple ocular and intracranial anomalies, genetic damage of

Congenital deafness, genetic damage of

Congenital diaphragmatic hernia, genetic damage of

Congenital erythropoietic porphyria, genetic damage of

Congenital facial diplegia, genetic damage of

Congenital fiber type disproportion, genetic damage of

Congenital gastrointestinal disorder, genetic damage of

Congenital generalized fibromatosis, genetic damage of

Congenital giant megaureter, genetic damage of

Congenital heart block, genetic damage of

Congenital heart disease ptosis hypodontia craniostosis, genetic damage of

Congenital heart disease radio ulnar synostosis mental retardation, genetic damage of

Congenital heart disorders, genetic damage of

Congenital heart septum defect, genetic damage of

Congenital hemidysplasia with ichtyosiform erythroderma and limbs defects, genetic damage of

Congenital hemolytic anemia, genetic damage of

Congenital hypomyelination neuropathy, genetic damage of

Congenital hypothyroidism, genetic damage of

Congenital hypotrichosis milia, genetic damage of

Congenital ichthyosis, genetic damage of

Congenital ichthyosis microcephalus quadriplegia, genetic damage of

Congenital ichtyosiform erythroderma, genetic damage of

Congenital kidney disorder, genetic damage of

Congenital lobar emphysema, genetic damage of

Congenital megacolon, genetic damage of

Congenital megalo-ureter, genetic damage of

Congenital mesoblastic nephroma, genetic damage of

Congenital microvillous atrophy, genetic damage of

Congenital mitral malformation, genetic damage of

Congenital mitral stenosis, genetic damage of

Congenital muscular dystrophy syringomyelia, genetic damage of

Congenital myopathy, genetic damage of

Congenital nephrotic syndrome, Finnish type, genetic damage of

Congenital nonhemolytic jaundice, genetic damage of

Congenital retinal telangiectasia, genetic damage of

Congenital short bowel, genetic damage of

Congenital short femur, genetic damage of

Congenital skeletal disorder, genetic damage of

Congenital skin disorder, genetic damage of

Congenital spherocytic anemia, genetic damage of

Congenital spherocytic hemolytic anemia, genetic damage of

Congenital stenosis of cervical medullary canal, genetic damage of

Congenital sucrose isomaltose malabsorption, genetic damage of

Congenital unilateral pulmonary hypoplasia, genetic damage of

Congenital vagal hyperreflexivity, genetic damage of

Congenital varicella syndrome, genetic damage of

Conn's syndrome, genetic damage of

Connective tissue dysplasia Spellacy type, genetic damage of

Connexin 26 anomaly, genetic damage of

Conotruncal heart malformations, genetic damage of

Conradi-Hünermann syndrome, genetic damage of

Constitutional growth delay, genetic damage of

Constrictive bronchiolitis, genetic damage of

Continuous muscle fiber activity hereditary, genetic damage of

Continuous spike-wave during slow sleep syndrome, genetic damage of

Contractural arachnodactyly, genetic damage of

Contractures ectodermal dysplasia cleft lip palate, genetic damage of

Contractures hyperkeratosis lethality, genetic damage of

Contractures of feet-muscle atrophy-oculomotor apraxia, genetic damage of

Conversion disorder, genetic damage of

Convulsions benign familial neonatal, genetic damage of

Convulsions benign familial neonatal dominant form, genetic damage of

Cooks syndrome, genetic damage of

Cooley's anemia, genetic damage of

Copper transport disease, genetic damage of

Coprastasophobia, genetic damage of (possible)

Coprophobia, genetic damage of (possible)

Coproporhyria, genetic damage of

Cor biloculare, genetic damage of

Cor triatriatum, genetic damage of

Cormier Rustin Munnich syndrome, genetic damage of

Corneal anesthesia deafness mental retardation, genetic damage of

Corneal cerebellar syndrome, genetic damage of

Corneal crystals myopathy neuropathy, genetic damage of

Corneal dystrophy, genetic damage of

Corneal dystrophy epithelial short stature, genetic damage of

Corneal dystrophy ichthyosis microcephaly mental retardation, genetic damage of

Corneal dystrophy perceptive deafness, genetic damage of

Corneal dystrophy pigmentary anomaly malabsorption, genetic damage of

Corneal endothelium dystrophy, genetic damage of

Cornelia de Lange syndrome, genetic damage of

Corneodermatoosseous syndrome, genetic damage of

Coronal dentin dysplasia, genetic damage of

Coronal synostosis syndactyly jejunal atresia, genetic damage of

Coronaro-cardiac fistula, genetic damage of

Coronary arteries congenital malformation, genetic damage of

Coronary artery aneurysm, genetic damage of

Corpus callosum agenesis, genetic damage of

Corpus callosum agenesis double urinary collecting, genetic damage of

Corpus callosum agenesis neuronopathy, genetic damage of

Corpus callosum agenesis of blepharophimosis Robin type, genetic damage of

Corpus callosum agenesis of with chorioretinal abnormalities, genetic damage of

Corpus callosum agenesis polysyndactyly, genetic damage of

Corpus callosum dysgenesis cleft spasm, genetic damage of

Corpus callosum dysgenesis hypopituitarism, genetic damage of

Corpus callosum dysgenesis X linked recessive, genetic damage of

Corrected transposition, genetic damage of

Corsello Opitz syndrome, genetic damage of

Cortada Koussef Matsumoto syndrome, genetic damage of

Cortes Lacassie syndrome, genetic damage of

Cortical blindness mental retardation polydactyly, genetic damage of

Cortical degeneration of the cerebellum parenchymatous, genetic damage of

Cortical hyperostosis syndactyly, genetic damage of

Corticobasal degeneration, genetic damage of

Costello syndrome, genetic damage of

Costocoracoid ligament congenitally short, genetic damage of

Costovertebral segmentation defect mesomelia, genetic damage of

Cote Adamopoulos Pantelakis syndrome, genetic damage of

Cote Katsantoni syndrome, genetic damage of

Cousin Walbraum Cegarra syndrome, genetic damage of

Covesdem syndrome, genetic damage of

Cowchock Wapner Kurtz syndrome, genetic damage of

Cowden's disease, genetic damage of

Coxoauricular syndrome, genetic damage of

Cramer Niederdellmann syndrome, genetic damage of

Cramp-fasciculations syndrome, genetic damage of

Crandall syndrome, genetic damage of

Crane Heise syndrome, genetic damage of

Cranio osteoarthropathy, genetic damage of

Cranioacrofacial syndrome, genetic damage of

Craniocerebellocardiac dysplasia, genetic damage of

Craniodiaphyseal dysplasia, genetic damage of

Craniodigital syndrome mental retardation, genetic damage of

Cranioectodermal dysplasia, genetic damage of

Craniofacial and osseous defects mental retardation, genetic damage of

Craniofacial and skeletal defects, genetic damage of

Craniofacial deafness hand syndrome, genetic damage of

Craniofacial dysostosis, genetic damage of

Craniofacial dysostosis arthrogryposis progeroid appearence, genetic damage of

Craniofacial dysynostosis, genetic damage of

Craniofaciocardioskeletal syndrome, genetic damage of

Craniofaciocervical osteoglyphic dysplasia, genetic damage of

Craniofrontonasal dysplasia, genetic damage of

Craniofrontonasal syndrome Teebi type, genetic damage of

Craniometaphyseal dysplasia, genetic damage of

Craniometaphyseal dysplasia dominant type, genetic damage of

Craniometaphyseal dysplasia recessive type, genetic damage of

Craniomicromelic syndrome, genetic damage of

Craniostenosis, genetic damage of

Craniostenosis cataract, genetic damage of

Craniostenosis with congenital heart disease mental retardation, genetic damage of

Craniosynostosis, genetic damage of

Craniosynostosis alopecia brain defect, genetic damage of

Craniosynostosis arthrogryposis cleft palate, genetic damage of

Craniosynostosis autosomal dominant, genetic damage of

Craniosynostosis brachydactyly, genetic damage of

Craniosynostosis cleft lip palate arthrogryposis, genetic damage of

Craniosynostosis contractures cleft, genetic damage of

Craniosynostosis Dandy Walker hydrocephalus, genetic damage of

Craniosynostosis exostoses nevus epibulbar dermoid, genetic damage of

Craniosynostosis fibular aplasia, genetic damage of

Craniosynostosis Fontaine type, genetic damage of

Craniosynostosis Maroteaux Fonfria type, genetic damage of

Craniosynostosis mental retardation clefting syndrome, genetic damage of

Craniosynostosis mental retardation heart defects, genetic damage of

Craniosynostosis Philadelphia type, genetic damage of

Craniosynostosis radial aplasia syndrome, genetic damage of

Craniosynostosis synostoses hypertensive nephropathy, genetic damage of

Craniosynostosis Warman type, genetic damage of

Craniotelencephalic dysplasia, genetic damage of

Crawfurd syndrome, genetic damage of

Creatine deficiency, genetic damage of

CREST syndrome, genetic damage of

Cretinism, genetic damage of

Cretinism athyreotic, genetic damage of

Cri du chat syndrome, genetic damage of

Crigler Najjar syndrome type I, genetic damage of

Crisponi syndrome, genetic damage of

Criss cross syndrome, genetic damage of

Criswick-Schepens syndrome, genetic damage of

Crohn's disease, genetic damage of

Crohn's disease of the esophagus, genetic damage of

Crome syndrome, genetic damage of

Cronkhite-Canada disease, genetic damage of

Crossed polydactyly type 1, genetic damage of

Crossed polysyndactyly, genetic damage of

Crouzon craniofacial dysostosis, genetic damage of

Crouzon disease, genetic damage of

Crow-Fukase syndrome, genetic damage of

Cryoglobulinemia, genetic damage of

Cryophobia, genetic damage of (possible)

Cryptogenic organized pneumopathy, genetic damage of

Cryptomicrotia brachydactyly syndrome, genetic damage of

Cryptomicrotia brachydactyly syndrome excess fingers, genetic damage of

Cryptophthalmos-syndactyly syndrome, genetic damage of

Cryptorchidism arachnodactyly mental retardation, genetic damage of

Cryroglobulinemia, genetic damage of

Crystal deposit disease, genetic damage of

Crystallophobia, genetic damage of (possible)

CTX, genetic damage of

Culler Jones syndrome, genetic damage of

Curly hair ankyloblepharon nail dysplasia syndrome, genetic damage of

Currarino triad, genetic damage of

Curry Hall syndrome, genetic damage of

Curth-Macklin type ichthyosis hystrix, genetic damage of

Curtis Rogers Stevenson syndrome, genetic damage of

Cushing syndrome, familial, genetic damage of

Cushing's symphalangism, genetic damage of

Cushing's syndrome, genetic damage of

Cutaneous lupus erythematosus, genetic damage of

Cutaneous photosensitivity colitis lethal, genetic damage of

Cutaneous T-cell lymphoma, genetic damage of

Cutaneous vascularitis, genetic damage of

Cutis Gyrata syndrome of Beare and Stevenson, genetic damage of

Cutis gyratum acanthosis nigricans craniosynostosis, genetic damage of

Cutis laxa, genetic damage of

Cutis laxa corneal clouding mental retardation, genetic damage of

Cutis laxa osteoporosis, genetic damage of

Cutis laxa with joint laxity and retarded development, genetic damage of

Cutis laxa, dominant type, genetic damage of

Cutis laxa, recessive type 1, genetic damage of

Cutis laxa, recessive type 2, genetic damage of

Cutis marmorata telangiectatica congenita, genetic damage of

Cutis marmorata telangiectatica congenita, genetic damage of

Cutis verticis gyrata, genetic damage of

Cutis verticis gyrata mental deficiency, genetic damage of

Cutis verticis gyrata thyroid aplasia mental retard, genetic damage of ation, genetic damage of

Cutler Bass Romshe syndrome, genetic damage of

Cyclic neutropenia, genetic damage of

Cyclic vomiting syndrome, genetic damage of

Cypress facial neuromusculoskeletal syndrome, genetic damage of

Cystathionine beta synthetase deficiency, genetic damage of

Cystic adenomatoid malformation of lung, genetic damage of

Cystic angiomatosis of bone, diffuse, genetic damage of

Cystic fibrosis, genetic damage of

Cystic fibrosis gastritis megaloblastic anemia, genetic damage of

Cystic hamartoma of lung and kidney, genetic damage of

Cystic hygroma, genetic damage of

Cystic hygroma lethal cleft palate, genetic damage of

Cystic medial necrosis of aorta, genetic damage of

Cystin transport, protein defect of, genetic damage of

Cystinosis, genetic damage of

Cystinuria, genetic damage of

Cystinuria-lysinuria, genetic damage of

Cytochrome C oxidase deficiency, genetic damage of

Cytomegalic inclusion disease, genetic damage of

Cytoplasmic body myopathy, genetic damage of

Czeizel Brooser syndrome, genetic damage of

Czeizel Losonci syndrome, genetic damage of

Czeizel syndrome, genetic damage of

D

D ercole syndrome, genetic damage of

D-glycerate dehydrogenase deficiency, genetic damage of

D-glycericacidemia, genetic damage of

D-minus hemolytic uremic syndrome (D-HUS), genetic damage of

D-plus hemolytic uremic syndrome (D+HUS), genetic damage of

Da Silva syndrome, genetic damage of

Daentl Towsend Siegel syndrome, genetic damage of

Dahlberg Borer Newcomer syndrome, genetic damage of

Daish Hardman Lamont syndrome, genetic damage of

Dandy Walker facial hemangioma, genetic damage of

Dandy Walker macrocephaly, genetic damage of

Dandy Walker malformation postaxial polydactyly, genetic damage of

Dandy Walker syndrome recessive form, genetic damage of

Dandy-Walker syndrome, genetic damage of

Daneman Davy Mancer syndrome, genetic damage of

Darier's disease, genetic damage of

Davenport Donlan syndrome, genetic damage of

David syndrome, genetic damage of

Davis Lafer syndrome, genetic damage of

De Barsy syndrome, genetic damage of

De Hauwere Leroy Adriaenssens syndrome, genetic damage of

De Morsier syndrome, genetic damage of

De Santis Cacchione syndrome, genetic damage of

Deaf blind hypopigmentation, genetic damage of

Deafness alopecia hypogonadism, genetic damage of

Deafness conductive ptosis skeletal anomalies, genetic damage of

Deafness conductive stapedial ear malformation facial palsy, genetic damage of

Deafness congenital onychodystrophy recessive, genetic damage of

Deafness craniofacial syndrome, genetic damage of

Deafness enamel hypoplasia nail defects, genetic damage of

Deafness epiphyseal dysplasia short stature, genetic damage of

Deafness goiter stippled epiphyses, genetic damage of

Deafness hyperuricemia neurologic ataxia, genetic damage of

Deafness hypogonadism syndrome, genetic damage of

Deafness hypospadias metacarpal and metatarsal syndrome, genetic damage of

Deafness mesenteric diverticula of small bowel neuropathy, genetic damage of

Deafness mixed with perilymphatic Gusher, X linked, genetic damage of

Deafness nephritis ano rectal malformation, genetic damage of

Deafness neurosensory pituitary dwarfism, genetic damage of

Deafness nonsyndromic, Connexin 26 linked, genetic damage of

Deafness oligodontia syndrome, genetic damage of

Deafness onychodystrophy dominant form, genetic damage of

Deafness peripheral neuropathy arterial disease, genetic damage of

Deafness progressive cataract autosomal dominant, genetic damage of

Deafness skeletal dysplasia lip granuloma, genetic damage of

Deafness symphalangism, genetic damage of

Deafness vitiligo achalasia, genetic damage of

Deafness white hair contractures papillomas, genetic damage of

Deafness X linked, DFN3, genetic damage of

Deafness, autosomal dominant nonsyndromic sensorineural, genetic damage of

Deafness, isolated, due to mitochondrial transmission, genetic damage of

Deafness, neurosensory nonsyndromic recessive, DFN, genetic damage of

Deafness, X linked, DFN, genetic damage of

Deal Barratt Dillon syndrome, genetic damage of

Deciduous skin, genetic damage of

Decompensated phoria, genetic damage of

Defect in synthesis of adenosylcobalamin, genetic damage of

Defective apolipoprotein B-100, genetic damage of

Defective expression of HLA class 2, genetic damage of

Degenerative motor system disease, genetic damage of

Degenerative optic myopathy, genetic damage of

Degos 'en cocarde' erythrokeratoderma, genetic damage of

Degos disease, genetic damage of

Dehydratase deficiency, genetic damage of

Dehydrated hereditary stomatocytosis, genetic damage of

Deipnophobia, genetic damage of (possible)

Dejerine-Sottas disease, genetic damage of

Delayed membranous cranial ossification, genetic damage of

Delayed speech facial asymmetry strabismus ear lobe creases, genetic damage of

Deletion 10p, genetic damage of

Deletion 10q, genetic damage of

Deletion 11p, genetic damage of

Deletion 11p 11p12, genetic damage of

Deletion 11p13, genetic damage of

Deletion 11q partial, genetic damage of

Deletion 12p12 p11, genetic damage of

Deletion 12p13, genetic damage of

Deletion 13q, genetic damage of

Deletion 13q14, genetic damage of

Deletion 13q22, genetic damage of

Deletion 13q32, genetic damage of

Deletion 14q partial duplication 14p partial, genetic damage of

Deletion 14q11, genetic damage of

Deletion 14q31, genetic damage of

Deletion 14qter, genetic damage of

Deletion 15q1, genetic damage of

Deletion 15q25, genetic damage of

Deletion 17q23 q24, genetic damage of

Deletion 18p, genetic damage of

Deletion 18q, genetic damage of

Deletion 18q23, genetic damage of

Deletion 1p, genetic damage of

Deletion 1p22 p13, genetic damage of

Deletion 1p31 p22, genetic damage of

Deletion 1p32, genetic damage of

Deletion 1p34 p32, genetic damage of

Deletion 1q21 q25, genetic damage of

Deletion 1q25 q32, genetic damage of

Deletion 1q32 q42, genetic damage of

Deletion 1q4, genetic damage of

Deletion 20p, genetic damage of

Deletion 21q22, genetic damage of

Deletion 2p22, genetic damage of

Deletion 2pter p24, genetic damage of

Deletion 2q, genetic damage of

Deletion 2q duplication 1p, genetic damage of

Deletion 2q24, genetic damage of

Deletion 3p, genetic damage of

Deletion 3p14 p11, genetic damage of

Deletion 3p25, genetic damage of

Deletion 3q13, genetic damage of

Deletion 3q21 23, genetic damage of

Deletion 3q27, genetic damage of

Deletion 4p, genetic damage of

Deletion 4p14 p16, genetic damage of

Deletion 4q, genetic damage of

Deletion 4q32, genetic damage of

Deletion 5q35, genetic damage of

Deletion 6p23, genetic damage of

Deletion 6q, genetic damage of

Deletion 6q1, genetic damage of

Deletion 6q13 q15, genetic damage of

Deletion 6q16 q21, genetic damage of

Deletion 6q2, genetic damage of

Deletion 7, genetic damage of

Deletion 7q2, genetic damage of

Deletion 7q21, genetic damage of

Deletion 7q3, genetic damage of

Deletion 8p, genetic damage of

Deletion 8p23 1, genetic damage of

Deletion 8q, genetic damage of

Deletion 8q12 21, genetic damage of

Deletion 8q21 q22, genetic damage of

Deletion 9p, genetic damage of

Deletion Xp22 pter, genetic damage of

Deletion Xq28, genetic damage of

Delleman oorthuys syndrome, genetic damage of

Delta-1-pyrroline-5-carboxylate dehydrogenase deficiency, genetic damage of

Delta-sarcoglycanopathy, genetic damage of

Dementia progressive lipomembranous polycysta, genetic damage of

Dementophobia, genetic damage of (possible)

Demonophobia, genetic damage of (possible)

Demyelinating diseases, genetic damage of

Dendrophobia, genetic damage of (possible)

Dennis Cohen syndrome, genetic damage of

Dennis Fairhurst Moore syndrome, genetic damage of

Dent disease, genetic damage of

Dental aberrations steroid dehydrogenase deficiency, genetic damage of

Dental tissue neoplasm, genetic damage of

Dentatorubral pallidoluysian atrophy, genetic damage of

Dentin dysplasia sclerotic bones, genetic damage of

Dentin dysplasia, coronal, genetic damage of

Dentin dysplasia, radicular, genetic damage of

Dentinogenesis imperfecta, genetic damage of

Dentophobia, genetic damage of (possible)

Depersonalization disorder, genetic damage of

Der kaloustian Jarudi Khoury syndrome, genetic damage of

Der Kaloustian McIntosh Silver syndrome, genetic damage of

Dercum disease, genetic damage of

Dermatitis herpetiformis, genetic damage of

Dermatocardioskeletal syndrome Boronne type, genetic damage of

Dermatographic uticaria, genetic damage of

Dermatoleukodystrophy, genetic damage of

Dermatomyositis, genetic damage of

Dermatoosteolysis Kirghizian type, genetic damage of

Dermatophobia, genetic damage of (possible)

Dermochondrocorneal dystrophy of François, genetic damage of

Dermoodontodysplasia, genetic damage of

Dermopathy restrictive lethal, genetic damage of

Desbuquois syndrome, genetic damage of

Desmin related myopathy, genetic damage of

Desmoid disease, genetic damage of

Desmoid tumor, genetic damage of

Desmoplastic small cell tumor, genetic damage of

Developmental delay hypotonia extremities hypertrophy, genetic damage of

Developmental dysphasia familial, genetic damage of

Devic syndrome, genetic damage of

Devriendt Legius Fryns syndrome, genetic damage of

Devriendt Vandenberghe Fryns syndrome, genetic damage of

Dexamethasone sensitive hypertension, genetic damage of

Dextrocardia, genetic damage of

Dextrocardia with situs inversus, genetic damage of

Dextrocardia-bronchiectasis-sinusitis, genetic damage of

Diabetes hypogonadism deafness mental retardation, genetic damage of

Diabetes insipidus, diabetes mellitus, optic atrophy, genetic damage of

Diabetes insipidus, nephrogenic type 1, genetic damage of

Diabetes insipidus, nephrogenic type 2, genetic damage of

Diabetes insipidus, nephrogenic type 3, genetic damage of

Diabetes insipidus, nephrogenic, dominant type, genetic damage of

Diabetes insipidus, nephrogenic, recessive type, genetic damage of

Diabetes mellitus, transient neonatal, genetic damage of

Diabetes persistent mullerian ducts, genetic damage of

Diabetes, insulin dependent, genetic damage of

Diabetic angiopathy, genetic damage of

Diabetic embryopathy, genetic damage of

Diabetic nephropathy, genetic damage of

Diabetic neuropathy, genetic damage of

Diamond Blackfan disease, genetic damage of

Diaphragmatic agenesia, genetic damage of

Diaphragmatic agenesis radial aplasia omphalocele, genetic damage of

Diaphragmatic defect limb deficiency skull defect, genetic damage of

Diaphragmatic hernia abnormal face limb, genetic damage of

Diaphragmatic hernia exomphalos corpus callosum agenesis, genetic damage of

Diaphragmatic hernia upper limb defects, genetic damage of

Diaphragmatic hernia, congenital, genetic damage of

Diarrhea chronic with villous atrophy, genetic damage of (possible)

Diarrhea polyendocrinopathy infections X linked, genetic damage of

Diastematomyelia, genetic damage of

Diastrophic dwarfism, genetic damage of

Diastrophic dysplasia, genetic damage of

Dibasic aminoaciduria 2, genetic damage of

Dibasic aminoaciduria type 1, genetic damage of

Dicarboxylicaminoaciduria, genetic damage of

Die Smulders Droog Van Dijk syndrome, genetic damage of

Die Smulders Vles Fryns syndrome, genetic damage of

Diencephalic syndrome, genetic damage of

Dieterich's disease, genetic damage of

Diffuse idiopathic skeletal hyperostosis, genetic damage of

Diffuse leiomyomatosis with Alport syndrome, genetic damage of

Diffuse neonatal haemangiomatosis, genetic damage of

Diffuse palmoplantar keratoderma Bothnian type, genetic damage of

DiGeorge syndrome, genetic damage of

Digestive duplication, genetic damage of

Digitorenocerebral syndrome, genetic damage of

Dihydropteridine reductase deficiency, genetic damage of

Dihydropyrimidine dehydrogenase deficiency, genetic damage of

Dilated cardiomyopathy, genetic damage of

Dimitri Sturge Weber syndrome, genetic damage of

Dincsoy Salih Patel syndrome, genetic damage of

Dinno Shearer Weisskopf syndrome, genetic damage of

Dinophobia, genetic damage of (possible)

Diomedi Bernardi Placidi syndrome, genetic damage of

Dionisi Vici Sabetta Gambarara syndrome, genetic damage of

Diphallia, genetic damage of

Diphallus rachischisis imperforate anus, genetic damage of

Diphosphoglycerate mutase deficiency of erythrocyte, genetic damage of

Diplophobia, genetic damage of (possible)

Dipsophobia, genetic damage of (possible)

Discoid lupus erythematosus, genetic damage of

Dislocation of the hip dysmorphism, genetic damage of

Disomy 1q12 q21, genetic damage of

Disomy 9q21, genetic damage of

Disorder in the hormonal synthesis with or without goiter, genetic damage of

Disorganization syndrome, genetic damage of

Dissecting cellulitis of the scalp, genetic damage of

Dissociative hysteria, genetic damage of

Distal arthrogryposis Moore Weaver type, genetic damage of

Distal myopathy, genetic damage of

Distal myopathy Markesbery-Griggs type, genetic damage of

Distal myopathy with vocal cord weakness, genetic damage of

Distal myopathy, Nonaka type, genetic damage of

Distal primary acidosis, familial, genetic damage of

Distichiasis heart congenital anomalies, genetic damage of

DK phocomelia syndrome, genetic damage of

Dobrow syndrome, genetic damage of

Dominant cleft palate, genetic damage of

Dominant ichthyosis vulgaris, genetic damage of

Dominant zonular cataract, genetic damage of

Donnai Barrow syndrome, genetic damage of

Donohue syndrome, genetic damage of

Door syndrome, genetic damage of

Dopa-responsive dystonia, genetic damage of

Dopamine beta-hydroxylase deficiency, genetic damage of

Doraphobia, genetic damage of (possible)

Double cortex, genetic damage of

Double discordia, genetic damage of

Double fingernail of fifth finger, genetic damage of

Double outlet left ventricle, genetic damage of

Double outlet right ventricle, genetic damage of

Double tachycardia induced by catecholamines, genetic damage of

Double uterus-hemivagina-renal agenesis, genetic damage of

Downs syndrome, genetic damage of

Doxorubicin-induced cardiomyopathy, genetic damage of

Doyne honeycomb retinal dystrophy, genetic damage of

Drachtman Weinblatt Sitarz syndrome, genetic damage of

Duane anomaly mental retardation, genetic damage of

Duane syndrome, genetic damage of

Dubin-Johnson syndrome, genetic damage of

Dubowitz syndrome, genetic damage of

Duchenne muscular dystrophy, genetic damage of

Ductular hepatic hypoplasia, genetic damage of

Duhring Brocq disease, genetic damage of

Duhring's disease, genetic damage of

Duker Weiss Siber syndrome, genetic damage of

Duodenal atresia, genetic damage of

Duodenal atresia tetralogy of Fallot, genetic damage of

Duplication 10p, genetic damage of

Duplication 10pter p13, genetic damage of

Duplication 10q partial, genetic damage of

Duplication 11q, genetic damage of

Duplication 11q23, genetic damage of

Duplication 12p, genetic damage of

Duplication 12q, genetic damage of

Duplication 13, genetic damage of

Duplication 13p, genetic damage of

Duplication 14q partial deletion 14p partial, genetic damage of

Duplication 14q prox, genetic damage of

Duplication 14q ter, genetic damage of

Duplication 15q, genetic damage of

Duplication 16p, genetic damage of

Duplication 16q, genetic damage of

Duplication 17p, genetic damage of

Duplication 17p11.2, genetic damage of

Duplication 18, genetic damage of

Duplication 18p, genetic damage of

Duplication 18q, genetic damage of

Duplication 19q, genetic damage of

Duplication 1p21 p32, genetic damage of

Duplication 1q12 q21, genetic damage of

Duplication 1q32 qter, genetic damage of

Duplication 1q42 qter, genetic damage of

Duplication 1q42.11 q42.12, genetic damage of

Duplication 20p, genetic damage of

Duplication 22, genetic damage of

Duplication 22q11 q13, genetic damage of

Duplication 2p, genetic damage of

Duplication 2p13 p21, genetic damage of

Duplication 2pter p24, genetic damage of

Duplication 2q, genetic damage of

Duplication 2q37, genetic damage of

Duplication 3p, genetic damage of

Duplication 3p25, genetic damage of

Duplication 3q, genetic damage of

Duplication 3q13.2 q25, genetic damage of

Duplication 4p, genetic damage of

Duplication 4q, genetic damage of

Duplication 4q21, genetic damage of

Duplication 4q25 qter, genetic damage of

Duplication 5pter p13.3, genetic damage of

Duplication 5q, genetic damage of

Duplication 6p, genetic damage of

Duplication 6q, genetic damage of

Duplication 7p, genetic damage of

Duplication 7p13 p12.2, genetic damage of

Duplication 7q, genetic damage of

Duplication 8p, genetic damage of

Duplication 8q, genetic damage of

Duplication 9p partial, genetic damage of

Duplication 9q21, genetic damage of

Duplication 9q32, genetic damage of

Duplication of leg mirror foot, genetic damage of

Duplication of the thumb unilateral biphalangeal, genetic damage of

Duplication of urethra, genetic damage of

Duplication Xp3, genetic damage of

Duplication Xpter Xq13, genetic damage of

Duplication Xq, genetic damage of

Duplication Xq13 1 q21 1, genetic damage of

Duplication Xq25, genetic damage of

Dupont Sellier Chochillon syndrome, genetic damage of

Dupuytren's contracture, genetic damage of

Dwarfism, genetic damage of

Dwarfism bluish sclerae, genetic damage of

Dwarfism deafness retinitis pigmentosa, genetic damage of

Dwarfism lethal type advanced bone age, genetic damage of

Dwarfism mental retardation eye abnormality, genetic damage of

Dwarfism short limb absent fibulas very short digits, genetic damage of

Dwarfism stiff joint ocular abnormalities, genetic damage of

Dwarfism syndesmodysplasic, genetic damage of

Dwarfism tall vertebrae, genetic damage of

Dwarfism thanatophoric, genetic damage of

Dwarfism thin bones multiple fractures, genetic damage of

Dyggve-Melchior-Clausen syndrome, genetic damage of

Dykes Markes Harper syndrome, genetic damage of

Dysautonomia (does not have to be familial), genetic damage of

Dyschondrosteosis, genetic damage of

Dyschondrosteosis nephritis, genetic damage of

Dyschromatosis universalis, genetic damage of

Dysencephalia splachnocystica or Meckel Gruber, genetic damage of

Dysequilibrium syndrome, genetic damage of

Dyserythropoietic anemia, congenital, genetic damage of

Dyserythropoietic anemia, congenital type 1, genetic damage of

Dyserythropoietic anemia, congenital type 2, genetic damage of

Dyserythropoietic anemia, congenital type 3, genetic damage of

Dysferlinopathy, genetic damage of

Dysfibrinogenemia, familial, genetic damage of

Dysgerminoma, genetic damage of

Dysharmonic skeletal maturation muscular fiber disproportion, genetic damage of

Dyskeratosis congenital syndrome, genetic damage of

Dyskeratosis congenita of Zinsser Cole Engman, genetic damage of

Dyskeratosis follicularis, genetic damage of

Dysmorphism abnormal vocalization mental retardation, genetic damage of

Dysmorphism cleft palate loose skin, genetic damage of

Dysmorphism multiple structural anomalies, genetic damage of

Dysmorphophobia, genetic damage of (possible)

Dysosteosclerosis, genetic damage of

Dysostosis, genetic damage of

Dysostosis acral with facial and genital abnormalities, genetic damage of

Dysostosis acrofacial postaxial, genetic damage of

Dysostosis peripheral, genetic damage of

Dysostosis Stanescu type, genetic damage of

Dysphasic dementia, hereditary, genetic damage of

Dysphonia, chronic spasmodic, genetic damage of

Dysplasia, genetic damage of

Dysplasia epiphysealis hemimelica, genetic damage of

Dysplasia faciogenital, genetic damage of

Dysplasia olfactogenitalis of de Morsier, genetic damage of

Dysplastic cortical hyperostosis, genetic damage of

Dysplastic nevus syndrome, genetic damage of

Dysprothrombinemia, genetic damage of

Dysraphism cleft lip palate limb reduction defects, genetic damage of

Dyssegmental dysplasia glaucoma, genetic damage of

Dyssegmental dysplasia Silverman Handmaker type, genetic damage of

Dysthymia, genetic damage of

Dystonia, genetic damage of

Dystonia musculorum deformans, genetic damage of

Dystonia musculorum deformans type 1, genetic damage of

Dystonia musculorum deformans type 2, genetic damage of

Dystonia progressive with diurnal variation, genetic damage of

Dystrophic epidermolysis bullosa, genetic damage of

Dystrophinopathy, genetic damage of

Dystrophy, myotonic, genetic damage of

Dystychiphobia, genetic damage of (possible)

E

EAF, genetic damage of

Eales disease, genetic damage of

Ear patella short stature syndrome, genetic damage of

Earlobes thickened conductive deafness, genetic damage of

Early infantile autism, genetic damage of

Eaton-Lambert syndrome, genetic damage of

Ebstein's anomaly, genetic damage of

Eccentrochondrodysplasia, genetic damage of

Eccrine acrospiroma, genetic damage of

Eclampsia, genetic damage of

Ecp syndrome, genetic damage of

Ectodermal dysplasia, genetic damage of

Ectodermal dysplasia absent dermatoglyphics, genetic damage of

Ectodermal dysplasia adrenal cyst, genetic damage of

Ectodermal dysplasia alopecia preaxial polydactyly, genetic damage of

Ectodermal dysplasia anhidrotic, genetic damage of

Ectodermal dysplasia arthrogryposis diabetes mellitus, genetic damage of

Ectodermal dysplasia Bartalos type, genetic damage of

Ectodermal dysplasia Berlin type, genetic damage of

Ectodermal dysplasia blindness, genetic damage of

Ectodermal dysplasia ectrodactyly macular dystrophy, genetic damage of

Ectodermal dysplasia hypohidrotic autosomal dominant, genetic damage of

Ectodermal dysplasia hypohidrotic hypothyroidism ciliary diskinesia, genetic damage of

Ectodermal dysplasia Margarita type, genetic damage of

Ectodermal dysplasia mental retardation CNS malformation, genetic damage of

Ectodermal dysplasia mental retardation syndactyly, genetic damage of

Ectodermal dysplasia neurosensory deafness, genetic damage of

Ectodermal dysplasia osteosclerosis, genetic damage of

Ectodermal dysplasia tricho odonto onychial type, genetic damage of

Ectodermal dysplasia, hydrotic, genetic damage of

Ectodermal dysplasia, hypohidrotic, autosomal recessive, genetic damage of

Ectodermal dysplasias, genetic damage of

Ectodermic dysplasia anhidrotic cleft lip, genetic damage of

Ectopia lentis chorioretinal dystrophy myopia, genetic damage of

Ectopia lentis isolated, genetic damage of

Ectopic coarctation, genetic damage of

Ectopic ossification familial type, genetic damage of

Ectopic pregnancy, genetic damage of

Ectrodactyly, genetic damage of

Ectrodactyly cardiopathy dysmorphism, genetic damage of

Ectrodactyly cleft palate syndrome, genetic damage of

Ectrodactyly diaphragmatic hernia corpus callosum, genetic damage of

Ectrodactyly dominant form, genetic damage of

Ectrodactyly ectrodermal dysplasia, genetic damage of

Ectrodactyly polydactyly, genetic damage of

Ectrodactyly recessive form, genetic damage of

Ectrodactyly spina bifida cardiopathy, genetic damage of

Ectrodactyly-ectodermal dysplasia-cleft lip/cleft palate, genetic damage of

Ectropion inferior cleft lip and or palate, genetic damage of

Edinburgh malformation syndrome, genetic damage of

Edwards Patton Dilly syndrome, genetic damage of

Edwards syndrome, genetic damage of

Eec syndrome, genetic damage of

Eec syndrome without cleft lip palate, genetic damage of

EEM syndrome, genetic damage of

Ehlers-Danlos syndrome, genetic damage of

Ehlers-Danlos syndrome type 1, genetic damage of

Ehlers-Danlos syndrome type 2, genetic damage of

Ehlers-Danlos syndrome type 3, genetic damage of

Ehlers-Danlos syndrome type 4, autosomal dominant, genetic damage of

Ehlers-Danlos syndrome type 6, genetic damage of

Ehlers-Danlos syndrome type 7A, genetic damage of

Ehlers-Danlos syndrome type 7B, genetic damage of

Ehlers-Danlos syndrome type 7C, genetic damage of

Ehlers-Danlos syndrome, arthrochalasic type, genetic damage of

Ehlers-Danlos syndrome, classic type, genetic damage of

Ehlers-Danlos syndrome, dermatosparaxis type, genetic damage of

Ehlers-Danlos syndrome, hypermobile type, genetic damage of

Ehlers-Danlos syndrome, vascular type, genetic damage of

Eijkman's syndrome, genetic damage of

Eisenmenger syndrome, genetic damage of

Eisoptrophobia, genetic damage of (possible)

Elattoproteus in context of NF, genetic damage of

Elective mutism, genetic damage of

Electron transfer flavoprotein, deficiency of, genetic damage of

Electrophobia, genetic damage of (possible)

Elejalde syndrome, genetic damage of

Elephant man in context of NF, genetic damage of

Elephantiasis, genetic damage of

Elliott Ludman Teebi syndrome, genetic damage of

Ellis Yale Winter syndrome, genetic damage of

Ellis-Van Creveld syndrome, genetic damage of

Emerinopathy, genetic damage of

Emery Nelson syndrome, genetic damage of

Emery-Dreifuss muscular dystrophy, genetic damage of

Emery-Dreifuss muscular dystrophy, dominant type, genetic damage of

Emery-Dreifuss muscular dystrophy, X linked, genetic damage of

Emetophobia, genetic damage of (possible)

Emphysema, genetic damage of

Emphysema, congenital lobar, genetic damage of

Emphysema-penoscrotal web-deafness-mental retardation, genetic damage of

Empty sella syndrome, genetic damage of

Enamel hypoplasia cataract hydrocephaly, genetic damage of

Enamel renal syndrome, genetic damage of

Encephalo cranio cutaneous lipomatosis, genetic damage of

Encephalocele, genetic damage of

Encephalocele anencephaly, genetic damage of

Encephalocele anterior, genetic damage of

Encephalocele frontal, genetic damage of

Encephalomyelitis, genetic damage of (possible)

Encephalomyelitis, myalgic, genetic damage of (possible)

Encephalopathy intracerebral calcification retinal, genetic damage of

Encephalopathy progressive optic atrophy, genetic damage of

Encephalopathy subacute spongiform, Gerstmann-Stra, genetic damage of

Encephalopathy-basal ganglia-calcification, genetic damage of

Encephalophathy recurrent of childhood, genetic damage of

Enchondromatosis (benign), genetic damage of

Enchondromatosis dwarfism deafness, genetic damage of

Endocardial fibroelastosis, genetic damage of

Endocrinopathy, genetic damage of

Endometrial stromal sarcoma, genetic damage of

Endometriosis, genetic damage of

Endomyocardial fibroelastosis, genetic damage of

Endomyocardial fibrosis, genetic damage of

Enetophobia, genetic damage of (possible)

Eng Strom syndrome, genetic damage of

Engelhard Yatziv syndrome, genetic damage of

Englemann disease, genetic damage of

Enochlophobia, genetic damage of (possible)

Enolase deficiency, genetic damage of

Enolase deficiency type 1, genetic damage of

Enolase deficiency type 2, genetic damage of

Enolase deficiency type 3, genetic damage of

Enolase deficiency type 4, genetic damage of

Enteropathica, genetic damage of

Eosinophilic cystitis, genetic damage of

Eosinophilic fasciitis, genetic damage of

Eosinophilic gastroenteritis, genetic damage of

Eosinophilic granuloma, genetic damage of

Eosinophilic lymphogranuloma, genetic damage of

Eosinophilic synovitis, genetic damage of

Eosophobia, genetic damage of (possible)

Ependymoblastoma, genetic damage of

Ependymoma, genetic damage of

Epidermal nevus syndrome, genetic damage of

Epidermal nevus vitamin D resistant rickets, genetic damage of

Epidermodysplasia verruciformis, genetic damage of

Epidermoid carcinoma, genetic damage of

Epidermolysa bullosa simplex and limb girdle muscular dystrophy, genetic damage of

Epidermolysis bullosa, genetic damage of

Epidermolysis bullosa acquisita, genetic damage of

Epidermolysis bullosa dystrophica, Bart type, genetic damage of

Epidermolysis bullosa dystrophica, dominant type, genetic damage of

Epidermolysis bullosa dystrophica, Hallopeau-Sieme, genetic damage of

Epidermolysis bullosa herpetiformis, Dowling-Meara, genetic damage of

Epidermolysis bullosa intraepidermic, genetic damage of

Epidermolysis bullosa inversa dystrophica, genetic damage of

Epidermolysis bullosa of hands and feet, genetic damage of

Epidermolysis bullosa simplex with anodontia, hair, genetic damage of

Epidermolysis bullosa simplex, Cockayne-Touraine type, genetic damage of

Epidermolysis bullosa simplex, Koebner type, genetic damage of

Epidermolysis bullosa simplex, Ogna type, genetic damage of

Epidermolysis bullosa simplex, Weber-Cockayne type, genetic damage of

Epidermolysis bullosa, dermolytic, genetic damage of

Epidermolysis bullosa, generalized atrophic benign, genetic damage of

Epidermolysis bullosa, junctional, genetic damage of

Epidermolysis bullosa, junctional, Herlitz-Pearson, genetic damage of

Epidermolysis bullosa, junctional, with pyloric atrophy, genetic damage of

Epidermolysis bullosa, pretibial, genetic damage of

Epidermolytic hyperkeratosis, genetic damage of

Epidermolytic palmoplantar keratoderma Vorner type, genetic damage of

Epilepsy, genetic damage of

Epilepsy benign neonatal dominant form, genetic damage of

Epilepsy benign neonatal recessive form, genetic damage of

Epilepsy juvenile absence, genetic damage of

Epilepsy mental deterioration Finnish type, genetic damage of

Epilepsy microcephaly skeletal dysplasia, genetic damage of

Epilepsy occipital calcifications, genetic damage of

Epilepsy progressive myoclonic type 2, genetic damage of

Epilepsy telangiectasia, genetic damage of

Epilepsy with myoclono-astatic crisis, genetic damage of

Epilepsy, benign occipital, genetic damage of

Epilepsy, myoclonic progressive familial, genetic damage of

Epilepsy, nocturnal, frontal lobe type, genetic damage of

Epilepsy, partial, familial, genetic damage of

Epilepsy, progressive myoclonic type 1, genetic damage of

Epimerase deficiency, genetic damage of

Epimetaphyseal dysplasia cataract, genetic damage of

Epimetaphyseal skeletal dysplasia, genetic damage of

Epiphyseal dysplasia dysmorphism camptodactyly, genetic damage of

Epiphyseal dysplasia hearing loss dysmorphism, genetic damage of

Epiphyseal dysplasia multiple, genetic damage of

Epiphyseal stippling syndrome osteoclastic hyperplasia, genetic damage of

Epiphysealis hemimelica dysplasia, genetic damage of

Epistaxiophobia, genetic damage of (possible)

Epithelial-myoepithelial carcinoma, genetic damage of

Epitheliopathy (APMPPE), genetic damage of

Epitheliopathy, acute posterior multifocal placoid, genetic damage of

EPP (erythropoietic protoporphyria), genetic damage of

Epstein syndrome, genetic damage of

Equinophobia, genetic damage of (possible)

Erb's palsy, genetic damage of, genetic damage of

Erb-Duchenne palsy, genetic damage of

Erdheim disease, genetic damage of

Erdheim-Chester syndrome, genetic damage of

Ereuthrophobia, genetic damage of (possible)

Ergophobia, genetic damage of (possible)

Eronen Somer Gustafsson syndrome, genetic damage of

Erosive pustular dermatosis of the scalp, genetic damage of

Erythema multiforme, genetic damage of

Erythermalgia, genetic damage of

Erythroblastopenia, genetic damage of

Erythroderma desquamativa of Leiner, genetic damage of

Erythroderma lethal congenital, genetic damage of

Erythrokeratodermia ataxia, genetic damage of

Erythrokeratodermia progressive symmetrica ichthyosis, genetic damage of

Erythrokeratodermia symmetrica progressiva, genetic damage of

Erythrokeratodermia variabilis ichthyosis, genetic damage of

Erythrokeratodermia variabilis, Mendes da Costa type, genetic damage of

Erythrokeratodermia with ataxia, genetic damage of

Erythrokeratolysis hiemalis ichthyosis, genetic damage of

Erythromelalgia, genetic damage of

Erythroplakia, genetic damage of

Erythropoietic protoporphyria, genetic damage of

Escher Hirt syndrome, genetic damage of

Esophageal atresia, genetic damage of

Esophageal atresia associated anomalies, genetic damage of

Esophageal atresia coloboma talipes, genetic damage of

Esophageal disorder, genetic damage of

Esophageal duodenal atresia abnormalities of hands, genetic damage of

Esophageal neoplasm, genetic damage of

Esophageal varices, genetic damage of

Esotropia, genetic damage of

Essential hypertension, genetic damage of

Essential iris atrophy, genetic damage of

Essential mixed cryoglobulinemia, genetic damage of

Essential thrombocytopenia, genetic damage of

Essential thrombocytosis, genetic damage of

Esthesioneuroblastoma, genetic damage of

Ethylmalonic aciduria, genetic damage of

Euhidrotic ectodermal dysplasia, genetic damage of

Eunuchoidism familial, genetic damage of

Euphobia, genetic damage of (possible)

Evan's syndrome, genetic damage of

Ewing's sarcoma, genetic damage of

Exencephaly, genetic damage of

Exfoliative dermatitis, genetic damage of

Exner syndrome, genetic damage of

Exomphalos-macroglossia-gigantism syndrome, genetic damage of

Exostoses, genetic damage of

Exostoses anetodermia brachydactyly type E, genetic damage of

Exostoses, multiple, genetic damage of

Exostoses, multiple, type 1, genetic damage of

Exostoses, multiple, type 2, genetic damage of

Exostoses, multiple, type 3, genetic damage of

Exstrophy of the bladder, genetic damage of

Exstrophy of the bladder-epispadias, genetic damage of

Exsudative retinopathy familial, autosomal dominant, genetic damage of

Exsudative retinopathy familial, autosomal recessive, genetic damage of

Exsudative retinopathy familial, X linked, recessive, genetic damage of

Exsudative retinopathy, familial, genetic damage of

Extrapyramidal disorder, genetic damage of

Extrasystoles short stature hyperpigmentation microcephaly, genetic damage of

Eye defects arachnodactyly cardiopathy, genetic damage of

Eyebrows and eyelashes absence mental retardation, genetic damage of

Eyebrows duplication syndactyly, genetic damage of

F

Fabry's disease, genetic damage of

Faces syndrome, genetic damage of

Facial asymmetry temporal seizures, genetic damage of

Facial clefting corpus callosum agenesis, genetic damage of

Facial dysmorphism macrocephaly myopia Dandy Walker type, genetic damage of

Facial dysmorphism shawl scrotum joint laxity syndrome, genetic damage of

Facial paralysis, genetic damage of (possible)

Facies unusual arthrogryposis advanced skeletal malformations, genetic damage of

Facio digito genital syndrome recessive form, genetic damage of

Facio skeletal genital syndrome Rippberger type, genetic damage of

Facio thoraco genital syndrome, genetic damage of

Faciocardiomelic dysplasia lethal, genetic damage of

Faciocardiorenal syndrome, genetic damage of

Faciodigitogenital syndrome, genetic damage of

Faciooculoacousticorenal syndrome, genetic damage of

Facioscapulohumeral muscular dystrophy, genetic damage of

Faciothoracoskeletal syndrome, genetic damage of

Factor II deficiency, genetic damage of

Factor II deficiency, genetic damage of

Factor IX deficiency, genetic damage of

Factor V deficiency, genetic damage of

Factor V Leiden mutation, genetic damage of

Factor VII deficiency, genetic damage of

Factor VIII deficiency, genetic damage of

Factor X deficiency, genetic damage of

Factor X deficiency, congenital, genetic damage of

Factor XI deficiency, congenital, genetic damage of

Factor XII deficiency, genetic damage of

Factor XIII deficiency, genetic damage of

Factor XIII deficiency, congenital, genetic damage of

Fahr's disease, genetic damage of

Fairbank disease, genetic damage of

Fallot complex mental growth retardation, genetic damage of

Fallot tetralogy, genetic damage of

Familial adenomatous polyposis, genetic damage of

Familial amyloid polyneuropathy, genetic damage of

Familial aortic dissection, genetic damage of

Familial band heterotopia, genetic damage of

Familial chondrocalcinosis, genetic damage of

Familial deafness, genetic damage of (possible)

Familial dysautonomia, genetic damage of

Familial emphysema, genetic damage of

Familial erythrophagocytic lymphohistiocytosis, genetic damage of

Familial hyperchylomicronemia, genetic damage of

Familial hyperlipoproteinemia, genetic damage of

Familial hyperlipoproteinemia type I, genetic damage of

Familial hyperlipoproteinemia type III, genetic damage of

Familial hyperlipoproteinemia type IV, genetic damage of

Familial hypersensitivity pneumonitis, genetic damage of

Familial hypertension, genetic damage of (possible)

Familial hypopituitarism, genetic damage of

Familial hypothyroidism, genetic damage of

Familial intestinal polyatresia syndrome, genetic damage of

Familial nasal acilia, genetic damage of

Familial non-immune hyperthyroidism, genetic damage of

Familial opposable triphalangeal thumbs duplication, genetic damage of

Familial partial epilepsy with variable focus, genetic damage of

Familial periodic paralysis, genetic damage of

Familial polyposis, genetic damage of

Familial porencephaly, genetic damage of

Familial supernumerary nipples, genetic damage of

Familial symmetric lipomatosis, genetic damage of

Familial thyroglossal duct cyst, genetic damage of

Familial Treacher Collins syndrome, genetic damage of

Familial veinous malformations, genetic damage of

Familial ventricular tachycardia, genetic damage of

Familial visceral myopathy, genetic damage of

Fanconi anemia type 1, genetic damage of

Fanconi anemia type 2, genetic damage of

Fanconi anemia type 3, genetic damage of

Fanconi Bickel syndrome, genetic damage of

Fanconi ichthyosis dysmorphism, genetic damage of

Fanconi like syndrome, genetic damage of

Fanconi pancytopenia, genetic damage of

Fanconi syndrome, renal, with nephrocalcinosis and renal stones, genetic damage of

Fanconi's anemia, genetic damage of

Fara Chlupackova syndrome, genetic damage of

Farber's disease, genetic damage of

Fas deficiency, genetic damage of

Fatal familial insomnia, genetic damage of

Fatty Liver, genetic damage of

Faulk Epstein Jones syndrome, genetic damage of

Faye Petersen Ward Carey syndrome, genetic damage of

Fazio Londe syndrome, genetic damage of

Fealty syndrome, genetic damage of

Febrile seizure, genetic damage of

Fechtner syndrome, genetic damage of

Feigenbaum Bergeron Richardson syndrome, genetic damage of

Feigenbaum Bergeron syndrome, genetic damage of

Feingold syndrome, genetic damage of

Feingold Trainer syndrome, genetic damage of

Felty's syndrome, genetic damage of

Female pseudohermaphrodism, genetic damage of

Female pseudohermaphrodism Genuardi type, genetic damage of

Femoral facial syndrome, genetic damage of

Femur bifid monodactylous ectrodactyly, genetic damage of

Femur fibula ulna syndrome, genetic damage of

Fenton Wilkinson Toselano syndrome, genetic damage of

Ferlini Ragno Calzolari syndrome, genetic damage of

Fernhoff Blackston Oakley syndrome, genetic damage of

Ferrocalcinosis cerebro vascular, genetic damage of

Fetal akinesia syndrome X linked, genetic damage of

Fetal and neonatal alloimmune thrombocytopenia, genetic damage of

Fetal brain disruption sequence, genetic damage of

Fetal edema, genetic damage of

Fetal left ventricular aneurysm, genetic damage of

FG syndrome, genetic damage of

FGDY, genetic damage of

Fiber type disproportion, congenital, genetic damage of

Fibrinogen deficiency, congenital, genetic damage of

Fibrochondrogenesis, genetic damage of

Fibrofolliculomas with trichodiscomas and acrochordons, genetic damage of

Fibrolipomatosis, genetic damage of

Fibromatosis, genetic damage of

Fibromatosis gingival hypertrichosis, genetic damage of

Fibromatosis gingival progressive deafness, genetic damage of

Fibromatosis multiple non ossifying, genetic damage of

Fibromuscular dysplasia, genetic damage of

Fibromuscular dysplasia of arteries, genetic damage of

Fibromyalgia, genetic damage of

Fibrosarcoma, genetic damage of

Fibrosing alveolitis, genetic damage of

Fibrosis, genetic damage of

Fibrous dysplasia, genetic damage of

Fibrous dysplasia of bone, genetic damage of

Fibrousdysplasia ossificans progressiva, genetic damage of

Fibula aplasia complex brachydactyly, genetic damage of

Fibular aplasia ectrodactyly, genetic damage of

Fibular hypoplasia femoral bowing oligodactyly, genetic damage of

Fibular hypoplasia scapulo pelvic dysplasia absent, genetic damage of

Filippi syndrome, genetic damage of

Fine Lubinsky syndrome, genetic damage of

Fingerprints absence syndactyly milia, genetic damage of

Fingers absence, genetic damage of

Finnish congenital nephrosis, genetic damage of

Finnish lethal neonatal metabolic syndrome, genetic damage of

Finnish type amyloidosis, genetic damage of

Finucane Kurtz Scott syndrome, genetic damage of

Fish malodor syndrome, genetic damage of

Fish-eye disease, genetic damage of

Fissured tongue, genetic damage of

Fistulous vegetative verrucous hydradenoma, genetic damage of

Fitz-Hugh-Curtis syndrome, genetic damage of

Fitzsimmons Walson Mellor syndrome, genetic damage of

Fitzsimmons-Guilbert syndrome, genetic damage of

Fitzsimmons-McLachlan-Gilbert syndrome, genetic damage of

Flat foot, genetic damage of

Floating-harbor syndrome, genetic damage of

Florid cystic endosalpingiosis of the uterus, genetic damage of

Flotch syndrome, genetic damage of

Flynn Aird syndrome, genetic damage of

Focal agyria pachygyria, genetic damage of

Focal alopecia congenital megalencephaly, genetic damage of

Focal dermal hypoplasia, genetic damage of

Focal dystonia, genetic damage of

Focal or multifocal malformations in neuronal migration, genetic damage of

Foix Chavany Marie syndrome, genetic damage of

Follicular atrophoderma-basal cell carcinoma, genetic damage of

Follicular hamartoma alopecia cystic fibrosis, genetic damage of

Follicular ichthyosis, genetic damage of

Follicular lymphoma, genetic damage of

Follicular lymphoreticuloma, genetic damage of

Fontaine Farriaux Blanckaert syndrome, genetic damage of

Forbes Albright syndrome, genetic damage of

Forbes disease, genetic damage of

Forestier's disease, genetic damage of

Forney Robinson Pascoe syndrome, genetic damage of

Fountain syndrome, genetic damage of

Fourth phacomatosis, genetic damage of

Fowler Christmas Chapele syndrome, genetic damage of

Fox-Fordyce disease, genetic damage of

Fragile X syndrome, genetic damage of

Fragile X syndrome type 1, genetic damage of

Fragile X syndrome type 2, genetic damage of

Fragile X syndrome type 3, genetic damage of

Fragoso Cid Garcia Hernandez syndrome, genetic damage of

Franceschetti-Klein syndrome, genetic damage of

Francheschini Vardeu Guala syndrome, genetic damage of

Francois dyscephalic syndrome, genetic damage of

Franek Bocker kahlen syndrome, genetic damage of

Fraser Jequier Chen syndrome, genetic damage of

Fraser like syndrome, genetic damage of

Fraser syndrome, genetic damage of

Frasier syndrome, genetic damage of

FRAXA syndrome, genetic damage of

FRAXD, genetic damage of

FRAXE syndrome, genetic damage of

Free sialic acid storage disease, genetic damage of

Freeman-Sheldon syndrome, genetic damage of

Freiberg's disease, genetic damage of

Freire Maia odontotrichomelic syndrome, genetic damage of

Freire Maia Pinheiro Opitz syndrome, genetic damage of

Frenkel Russe syndrome, genetic damage of

Frey's syndrome, genetic damage of

Frias syndrome, genetic damage of

Fried Goldberg Mundel syndrome, genetic damage of

Friedel Heid Grosshans syndrome, genetic damage of

Friedman Goodman syndrome, genetic damage of

Friedreich ataxia congenital glaucoma, genetic damage of

Friedreich's ataxia, genetic damage of

Frigophobia, genetic damage of (possible)

Froelich's syndrome, genetic damage of

Frölich's syndrome, genetic damage of

Fronto nasal malformation cloacal exstrophy, genetic damage of

Fronto-facio-nasal dysplasia, genetic damage of

Frontofacionasal dysplasia type Al gazali, genetic damage of

Frontometaphyseal dysplasia, genetic damage of

Frontonasal dysplasia, genetic damage of

Frontonasal dysplasia acromelic, genetic damage of

Frontonasal dysplasia klippel feil syndrome, genetic damage of

Frontonasal dysplasia phocomelic upper limbs, genetic damage of

Frontotemporal lobe dementia, genetic damage of

Froster huch syndrome, genetic damage of

Froster Iskenius Waterson syndrome, genetic damage of

Fructose intolerance, genetic damage of

Fructose-1,6-bisphosphatase deficiency, genetic damage of

Fructose-1-phosphate aldolase deficiency, hereditary, genetic damage of

Fructosemia, hereditary, genetic damage of

Fructosuria, genetic damage of

Frydman Cohen Ashenazi syndrome, genetic damage of

Frydman Cohen Karmon syndrome, genetic damage of

Fryer syndrome, genetic damage of

Fryns Fabry Remans syndrome, genetic damage of

Fryns Hofkens Fabry syndrome, genetic damage of

Fryns Smeets Thiry syndrome, genetic damage of

Fryns syndrome, genetic damage of

Fucosidosis, genetic damage of

Fucosidosis type 1, genetic damage of

Fuhrmann Rieger De sousa syndrome, genetic damage of

Fukuda Miyanomae Nakata syndrome, genetic damage of

Fukuyama type muscular dystrophy, genetic damage of

Fumarase deficiency, genetic damage of

Fumaric aciduria, genetic damage of

Fumarylacetoacetase deficiency, genetic damage of

Functioning pancreatic endocrine tumor, genetic damage of

Fuqua Berkovitz syndrome, genetic damage of

Furlong Kurczynski Hennessy syndrome, genetic damage of

Furukawa Takagi Nakao syndrome, genetic damage of

Fused mandibular incisors, genetic damage of

G

G syndrome, genetic damage of

Galactocerebrosidase deficiency, genetic damage of

Galactokinase deficiency, genetic damage of

Galactosamine-6-sulfatase deficiency, genetic damage of

Galactose-1-phosphate uridyltransferase deficiency, genetic damage of

Galactosemia, genetic damage of

Galactosialidosis, genetic damage of

Galloway Mowat syndrome, genetic damage of

Gamborg Nielsen syndrome, genetic damage of

Game Friedman Paradice syndrome, genetic damage of

Gamma aminobutyric acid transaminase deficiency, genetic damage of

Gamma-cystathionase deficiency, genetic damage of

Gamma-sarcoglycanopathy, genetic damage of

Gamophobia, genetic damage of (possible)

Gamstorp episodic adynamy, genetic damage of

Ganglioglioma, genetic damage of

Gangliosidosis (Type2)(GM2), genetic damage of

Gangliosidosis GM1 type 3, genetic damage of

Gangliosidosis type1, genetic damage of

Gapo syndrome, genetic damage of

Garcia Torres Guarner syndrome, genetic damage of

Gardner Morrisson Abbot syndrome, genetic damage of

Gardner Silengo Wachtel syndrome, genetic damage of

Gardner's syndrome, genetic damage of

Gardner-Diamond syndrome, genetic damage of

Garret Tripp syndrome, genetic damage of

Gastric lymphoma, genetic damage of

Gastritis, familial giant hypertrophic, genetic damage of

Gastro-enteropancreatic neuroendocrine tumor, genetic damage of

Gastrocutaneous syndrome, genetic damage of

Gastroenteritis, eosinophilic, genetic damage of

Gastroesophageal reflux, genetic damage of

Gastrointestinal autonomic nerve tumor, genetic damage of

Gastrointestinal neoplasm, genetic damage of

Gaucher Disease, genetic damage of

Gaucher disease type 1, genetic damage of

Gaucher disease type 2, genetic damage of

Gaucher disease type 3, genetic damage of

Gaucher ichthyosis restrictive dermopathy, genetic damage of

Gaucher-like disease, genetic damage of

Gay Feinmesser Cohen syndrome, genetic damage of

Geen Sandford Davison syndrome, genetic damage of

Gelatinous ascites, genetic damage of

Geleophysic dwarfism, genetic damage of

Gelineau disease, genetic damage of

Geliphobia, genetic damage of (possible)

Gemignani syndrome, genetic damage of

Gemss syndrome, genetic damage of

Generalized malformations in neuronal migration, genetic damage of

Generalized resistance to thyroid hormone, genetic damage of

Generalized seizure, genetic damage of

Generalized torsion dystonia, genetic damage of

Genes syndrome, genetic damage of

Genetic reflex epilepsy, genetic damage of

Geniophobia, genetic damage of (possible)

Genital anomaly cardiomyopathy, genetic damage of

Genital dwarfism, genetic damage of

Genital dwarfism turner type, genetic damage of

Genito palatocardiac syndrome, genetic damage of

Genuphobia, genetic damage of (possible)

Geographic tongue, genetic damage of

Gerascophobia, genetic damage of (possible)

Gerhardt syndrome, genetic damage of

German syndrome, genetic damage of

Gerodermia osteodysplastica, genetic damage of

Gershinibaruch Leibo syndrome, genetic damage of

Gerstmann syndrome, genetic damage of

Gestational diabetes mellitus, genetic damage of

Gestational pemphigoid, genetic damage of

Gestational trophoblastic disease, genetic damage of

Ghosal syndrome, genetic damage of

Ghose Sachdev Kumar syndrome, genetic damage of

Gianotti-Crosti syndrome, genetic damage of

Giant axonal neuropathy, genetic damage of

Giant cell arteritis, genetic damage of

Giant cell myocarditis, genetic damage of

Giant congenital nevi, genetic damage of

Giant ganglionic hyperplasia, genetic damage of

Giant hypertrophic gastritis, genetic damage of

Giant mammary hamartoma, genetic damage of

Giant pigmented hairy nevus, genetic damage of

Giant platelet syndrome, genetic damage of

Gigantism, genetic damage of

Gigantism advanced bone age hoarse cry, genetic damage of

Gigantism exomphalos macroglossia, genetic damage of

Gigantism partial, nevi, hemihypertrophy, macrocephaly, genetic damage of

Gilbert's syndrome, genetic damage of

Gilles de la Tourette's syndrome, genetic damage of

Gingival fibromatosis dominant, genetic damage of

Gingival fibromatosis facial dysmorphism, genetic damage of

Gingival fibrosis, genetic damage of

Gingivitis, genetic damage of

Girate atrophy of choroid and retina, genetic damage of

Glanzmann thrombasthenia, genetic damage of

Glass Chapman Hockley syndrome, genetic damage of

Glastre Cochat Bouvier syndrome, genetic damage of

Glaucoma ecopia microspherophakia stiff joints short stature, genetic damage of

Glaucoma iridogoniodysgenesia, genetic damage of

Glaucoma sleep apnea, genetic damage of

Glaucoma type 1C, genetic damage of

Glaucoma, congenital, genetic damage of

Glaucoma, hereditary, genetic damage of

Glaucoma, hereditary adult type 1A, genetic damage of

Glaucoma, hereditary juvenile type 1B, genetic damage of

Glaucoma, primary infantile type 3A, genetic damage of

Glaucoma, primary infantile type 3B, genetic damage of

Glioblastoma, genetic damage of

Glioblastoma multiforme, genetic damage of

Glioma, genetic damage of

Gliomatosis cerebri, genetic damage of

Gliosarcoma, genetic damage of

Glomerulonephritis sparse hair telangiectases, genetic damage of

Glomerulosclerosis, genetic damage of

Gloomy face syndrome, genetic damage of

Glossodynia, genetic damage of

Glossopalatine ankylosis cataracts digital anomalies, genetic damage of

Glossopalatine ankylosis micrognathia ear anomalies, genetic damage of

Glossopharyngeal neuralgia, genetic damage of

Glossophobia, genetic damage of (possible)

Glucagonoma, genetic damage of

Glucocorticoid deficiency, familial, genetic damage of

Glucocorticoid resistance, genetic damage of

Glucocorticoid sensitive hypertension, genetic damage of

Glucose 6 phosphate dehydrogenase deficiency, genetic damage of

Glucose-6-phosphate translocase deficiency, genetic damage of

Glucose-galactose malabsorption, genetic damage of

Glucosephosphate isomerase deficiency, genetic damage of

Glucosidase acid-1,4-alpha deficiency, genetic damage of

Glut2 deficiency, genetic damage of

Glutamate decarboxylase deficiency, genetic damage of

Glutamate-aspartate transport defect, genetic damage of

Glutaricaciduria I, genetic damage of

Glutaricaciduria II, genetic damage of

Glutaryl-CoA dehydrogenase deficiency, genetic damage of

Glyceraldehyde-3-phosphate dehydrogenase deficiency, genetic damage of

Glycerol kinase deficiency, genetic damage of

Glycine synthase deficiency, genetic damage of

Glycinemia, ketotic, genetic damage of

Glycogen storage disease type 1A, genetic damage of

Glycogen storage disease type 1B, genetic damage of

Glycogen storage disease type 1C, genetic damage of

Glycogen storage disease type 1D, genetic damage of

Glycogen storage disease type 6, due to phosphorylation, genetic damage of

Glycogen storage disease type 7, genetic damage of

Glycogen storage disease type 9, genetic damage of

Glycogen storage disease type I, genetic damage of

Glycogen storage disease type II, genetic damage of

Glycogen storage disease type V, genetic damage of

Glycogen storage disease type VI, genetic damage of

Glycogen storage disease type VII, genetic damage of

Glycogen storage disease type VIII, genetic damage of

Glycogenosis type II, genetic damage of

Glycogenosis type III, genetic damage of

Glycogenosis type IV, genetic damage of

Glycogenosis type V, genetic damage of

Glycogenosis type VI, genetic damage of

Glycogenosis type VII, genetic damage of

Glycogenosis type VIII, genetic damage of

Glycogenosis, type 0, genetic damage of

Glycosuria, genetic damage of

GM2 gangliosidosis, 0 variant, genetic damage of

GM2-gangliosidosis, B, B1, AB variant, genetic damage of

GMS syndrome, genetic damage of

Goiter-deafness syndrome, genetic damage of

Goldberg Bull syndrome, genetic damage of

Goldberg syndrome, genetic damage of

Goldblatt Wallis syndrome, genetic damage of

Goldblatt Wallis Zieff syndrome, genetic damage of

Goldblatt Viljoen syndrome, genetic damage of

Goldenhar disease, genetic damage of

Goldskag Cooks Hertz syndrome, genetic damage of

Goldstein Hutt syndrome, genetic damage of

Gollop Coates syndrome, genetic damage of

Gollop syndrome, genetic damage of

Gollop Wolfgang complex, genetic damage of

Goltz syndrome, genetic damage of

Gombo syndrome, genetic damage of

Gomez and López-Hernández syndrome, genetic damage of

Gonadal dysgenesis, genetic damage of

Gonadal dysgenesis mixed, genetic damage of

Gonadal dysgenesis Turner type, genetic damage of

Gonadal dysgenesis XY type associated anomalies, genetic damage of

Gonadal dysgenesis, XX type, genetic damage of

Gonadal dysgenesis, XY female type, genetic damage of

Goniodysgenesis mental retardation short stature, genetic damage of

Gonococcal conjunctivitis, genetic damage of

Gonodal dysgenesis XX type deafness, genetic damage of

Gonzales Del Angel syndrome, genetic damage of

Goodman camptodactyly, genetic damage of

Goodman syndrome, genetic damage of

Goodpasture pneumorenal syndrome, genetic damage of

Goodpasture's syndrome, genetic damage of

Gordon hyperkaliemia-hypertension syndrome, genetic damage of

Gordon syndrome, genetic damage of

Gorham syndrome, genetic damage of

Gorham-Stout disease, genetic damage of

Gorlin Bushkell Jensen syndrome, genetic damage of

Gorlin Chaudhry Moss syndrome, genetic damage of

Gottron's syndrome, genetic damage of

Gougerot-Sjogren syndrome, genetic damage of

Gouty nephropathy, familial, genetic damage of

Graft versus host disease, genetic damage of

Graham Boyle Troxell syndrome, genetic damage of

Grand Kaine Fulling syndrome, genetic damage of

Grant syndrome, genetic damage of

Granulocytopenia, genetic damage of

Granuloma annulare, genetic damage of

Granulomatosis, lymphomatoid, genetic damage of

Granulomatous allergic angiitis, genetic damage of

Granulomatous hypophysitis, genetic damage of

Granulomatous rosacea, genetic damage of

Graves' disease, genetic damage of

Gray platelet syndrome, genetic damage of

Great vessels transposition, genetic damage of

Greenberg dysplasia, genetic damage of

Greig cephalopolysyndactyly syndrome, genetic damage of

Greig syndrome, genetic damage of

Griscelli disease, genetic damage of

Grix Blankenship Peterson syndrome, genetic damage of

Groll Hirschowitz syndrome, genetic damage of

Grosse syndrome, genetic damage of

Grover's disease, genetic damage of

Growth deficiency brachydactyly unusual facies, genetic damage of

Growth delay, constitutional, genetic damage of

Growth hormone deficiency, genetic damage of

Growth mental deficiency syndrome of Myhre, genetic damage of

Growth retardation alopecia pseudoanodontia optic, genetic damage of

Growth retardation hydrocephaly lung hypoplasia, genetic damage of

Growth retardation mental retardation phalangeal hypoplasia, genetic damage of

Grubben Decock Borghgraef syndrome, genetic damage of

GTP cyclohydrolase deficiency, genetic damage of

Guanidinoacetate methyltransferase deficiency, genetic damage of

Guibaud Vainsel syndrome, genetic damage of

Guillan-Barre syndrome, genetic damage of

Guizar Vasquez Luengas syndrome, genetic damage of

Guizar Vasquez Sanchez Manzano syndrome, genetic damage of

Gunal Seber Basaran syndrome, genetic damage of

Gupta Patton syndrome, genetic damage of

Gurrieri Sammito Bellussi syndrome, genetic damage of

Gusher syndrome, genetic damage of

Gymnophobia, genetic damage of (possible)

Gyrate atrophy, genetic damage of

Gyrate atrophy of the retina, genetic damage of

H

Haas Chir Robinson syndrome, genetic damage of

Haemangioendothelioma, genetic damage of

Haemorrhagic proctocolitis, genetic damage of

Hageman factor deficiency, genetic damage of

Hagemoser Weinstein Bresnick syndrome, genetic damage of

Hagiophobia, genetic damage of (possible)

Hailey-Hailey disease, genetic damage of

Hair defect photosensitivity mental retardation, genetic damage of

Hairy cell leukemia, genetic damage of

Hairy throat syndrome, genetic damage of

Hairy tongue, genetic damage of

Hajdu-Cheney syndrome, genetic damage of

Halal Setton Wang syndrome, genetic damage of

Halal syndrome, genetic damage of

Hall Riggs mental retardation syndrome, genetic damage of

Hallermann Streiff Francois syndrome, genetic damage of

Hallermann-Streiff syndrome, genetic damage of

Hallervorden-Spatz disease, genetic damage of

Hallux varus and preaxial polysyndactyly, genetic damage of

Hamanishi Ueba Tsuji syndrome, genetic damage of

Hamano Tsukamoto syndrome, genetic damage of

Hamartoma sebaceus of Jadassohn, genetic damage of

Hamman-Rich disease, genetic damage of

Hand and foot deformity flat facies, genetic damage of

Hand foot uterus syndrome, genetic damage of

Hand wringing Rett syndrome, genetic damage of

Hand-foot-mouth disease, genetic damage of

Hand-Schuller-Christian disease, genetic damage of

Hanhart syndrome, genetic damage of

Hapnes Boman Skeie syndrome, genetic damage of

Hard skin syndrome Parana type, genetic damage of

HARD syndrome, genetic damage of

Harding ataxia, genetic damage of

Harlequin type ichthyosis, genetic damage of

Harpaxophobia, genetic damage of (possible)

Harrod Doman Keele syndrome, genetic damage of

Hartnup disease, genetic damage of

Hartsfield Bixler Demyer syndrome, genetic damage of

Hashimoto struma, genetic damage of

Hashimoto's syndrome, genetic damage of

Hashimoto-Pritzker syndrome, genetic damage of

Haspeslagh Fryns Muelenaere syndrome, genetic damage of

Hay Wells syndrome recessive type, genetic damage of

Hay-Wells syndrome, genetic damage of

Headache, cluster, genetic damage of

Hearing disorder, genetic, genetic damage of

Hearing impairment, genetic, genetic damage of

Hearing loss, genetic, genetic damage of

Heart aneurysm, genetic damage of

Heart block, genetic damage of

Heart block progressive, familial, genetic damage of

Heart defect round face congenital retarded development, genetic damage of

Heart defect tongue hamartoma polysyndactyly, genetic damage of

Heart defects limb shortening, genetic damage of

Heart hand syndrome Spanish type, genetic damage of

Heart hypertrophy, hereditary, genetic damage of

Heart situs anomaly, genetic damage of

Heart tumor of the adult, genetic damage of

Heart tumor of the child, genetic damage of

HEC syndrome, genetic damage of

Hecht Scott syndrome, genetic damage of

Heckenlively syndrome, genetic damage of

Heide syndrome, genetic damage of

Heliophobia, genetic damage of (possible)

Helmerhorst Heaton Crossen syndrome, genetic damage of

HEM dysplasia, genetic damage of

Hemangioblastoma, genetic damage of

Hemangioma, genetic damage of

Hemangioma thrombocytopenia syndrome, genetic damage of

Hemangiomas cavernous of face supraumbilical midline raphe, genetic damage of

Hemangiopericytoma, genetic damage of

Hemeralopia, congenital essential, genetic damage of

Hemeralopia, familial, genetic damage of

Hemi 3 syndrome, genetic damage of

Hemifacial atrophy agenesis of the caudate nucleus, genetic damage of

Hemifacial atrophy progressive, genetic damage of

Hemifacial hyperplasia strabismus, genetic damage of

Hemifacial microsomia, genetic damage of

Hemihypertrophy in context of NF, genetic damage of

Hemihypertrophy intestinal web corneal opacity, genetic damage of

Hemimegalencephaly, genetic damage of

Hemiplegia, genetic damage of

Hemiplegic migraine, familial, genetic damage of

Hemoglobin C disease, genetic damage of

Hemoglobin E disease, genetic damage of

Hemoglobin SC disease, genetic damage of

Hemoglobinopathy, genetic damage of

Hemoglobinuria, genetic damage of

Hemolytic anemia lethal genital anomalies, genetic damage of

Hemolytic-uremic syndrome, genetic damage of

Hemophagocytic lymphohistiocytosis, genetic damage of

Hemophagocytic reticulosis (or reticulitis), genetic damage of

Hemophilia, genetic damage of

Hemophilic arthropathy, genetic damage of

Hemophobia, genetic damage of (possible)

Hemorrhagic thrombocythemia, genetic damage of

Hemorrhagiparous thrombocytic dystrophy, genetic damage of

Hemosiderosis, genetic damage of

Hennekam Beemer syndrome, genetic damage of

Hennekam Koss de Geest syndrome, genetic damage of

Hennekam syndrome, genetic damage of

Hennekam Van der Horst syndrome, genetic damage of

Heparane sulfamidase deficiency, genetic damage of

Hepatic cystic hamartoma, genetic damage of

Hepatic ductular hypoplasia, genetic damage of

Hepatic fibrosis, genetic damage of

Hepatic fibrosis renal cysts mental retardation, genetic damage of

Hepatic venoocclusive disease, genetic damage of

Hepatoblastoma, genetic damage of

Hepatocellular carcinoma, genetic damage of

Hepatorenal syndrome, genetic damage of

Hepatorenal tyrosinemia, genetic damage of

Hereditary amyloidosis, genetic damage of

Hereditary angioedema, genetic damage of

Hereditary ataxia, genetic damage of

Hereditary carnitine deficiency, genetic damage of

Hereditary carnitine deficiency myopathy, genetic damage of

Hereditary ceroid-lipofuscinosis, genetic damage of

Hereditary coproporphyria, genetic damage of

Hereditary deafness, genetic damage of

Hereditary elliptocytosis, genetic damage of

Hereditary fructose intolerance, genetic damage of

Hereditary hearing disorder, genetic damage of

Hereditary hearing loss, genetic damage of

Hereditary hemochromatosis, genetic damage of

Hereditary hemorrhagic telangiectasia, genetic damage of

Hereditary hyperuricemia, genetic damage of

Hereditary lymphedema, genetic damage of

Hereditary macrothrombocytopenia, genetic damage of

Hereditary methemoglobinemia, recessive, genetic damage of

Hereditary motor and sensory neuropathy, genetic damage of

Hereditary myopathy with intranuclear filamentous, genetic damage of

Hereditary nodular heterotopia, genetic damage of

Hereditary non-spherocytic hemolytic anemia, genetic damage of

Hereditary pancreatitis, genetic damage of

Hereditary paroxysmal cerebral ataxia, genetic damage of

Hereditary peripheral nervous disorder, genetic damage of

Hereditary primary Fanconi disease, genetic damage of

Hereditary resistance to anti-vitamin K, genetic damage of

Hereditary sensory and autonomic neuropathy 3, genetic damage of

Hereditary sensory and autonomic neuropathy 4, genetic damage of

Hereditary sensory neuropathy type I, genetic damage of

Hereditary sensory neuropathy type II, genetic damage of

Hereditary spastic paraplegia, genetic damage of

Hereditary spherocytic hemolytic anemia, genetic damage of

Hereditary spherocytosis, genetic damage of

Hereditary type 1 neuropathy, genetic damage of

Hereditary type 2 neuropathy, genetic damage of

Hereditary tyrosinemia, genetic damage of

Hermansky-Pudlak syndrome, genetic damage of

Hermaphroditism, genetic damage of

Hernandez Aguire Negrete syndrome, genetic damage of

Herpetophobia, genetic damage of (possible)

Herrmann Opitz arthrogryposis syndrome, genetic damage of

Herrmann Opitz craniosynostosis, genetic damage of

Hers disease, genetic damage of

Hersh Podruch Weisskopk syndrome, genetic damage of

Heterophobia, genetic damage of (possible)

Heterotaxia (generic term), genetic damage of

Heterotaxia autosomal dominant type, genetic damage of

Heterotaxy with polysplenia or asplenia, genetic damage of

Heterotaxy, visceral, X linked, genetic damage of

Hexosaminidase A deficiency, genetic damage of

Hexosaminidases A and B deficiency, genetic damage of

HHH syndrome, genetic damage of

Hidradenitis suppurativa, genetic damage of

Hidradenitis suppurativa familial, genetic damage of

Hidrotic ectodermal dysplasia type Christianson Fouris, genetic damage of

High scapula, genetic damage of

High-molecular-weight kininogen deficiency, congenital, genetic damage of

Hillig syndrome, genetic damage of

Hing Torack Dowston syndrome, genetic damage of

Hinson-Pepys disease, genetic damage of

Hip dislocation, genetic damage of

Hip Dysplasia, genetic damage of

Hip dysplasia Beukes type, genetic damage of

Hip luxation, genetic damage of

Hip subluxation, genetic damage of

Hipo syndrome, genetic damage of

Hippel Lindau disease, genetic damage of

Hirschsprung disease ganglioneuroblastoma, genetic damage of

Hirschsprung disease polydactyly heart disease, genetic damage of

Hirschsprung disease type 2, genetic damage of

Hirschsprung disease type 3, genetic damage of

Hirschsprung disease type d brachydactyly, genetic damage of

Hirschsprung disease with pigmentary anomaly, genetic damage of

Hirschsprung microcephaly cleft palate, genetic damage of

Hirschsprung nail hypoplasia dysmorphism, genetic damage of

Hirschsprung's disease, genetic damage of

Hirsutism congenital gingival hyperplasia, genetic damage of

Hirsutism skeletal dysplasia mental retardation, genetic damage of

His bundle tachycardia, genetic damage of

Histidase deficiency, genetic damage of

Histidinemia, genetic damage of

Histidinuria renal tubular defect, genetic damage of

Histiocytosis X, genetic damage of

Histiocytosis, Non-Langerhans-Cell, genetic damage of

Hittner Hirsch Kreh syndrome, genetic damage of

Hm syndrome, genetic damage of

HMC syndrome, genetic damage of

Hmg coa synthetase deficiency, genetic damage of

Hodgkin lymphoma, genetic damage of

Hodgkin's disease, genetic damage of

Hodophobia, genetic damage of (possible)

Hoepffner Dreyer Reimers syndrome, genetic damage of

Hollow visceral myopathy, genetic damage of

Holmes Benacerraf syndrome, genetic damage of

Holmes Borden syndrome, genetic damage of

Holmes Collins syndrome, genetic damage of

Holmes Gang syndrome, genetic damage of

Holoacardius amorphus, genetic damage of

Holocarboxylase synthetase deficiency, genetic damage of

Holoprosencephaly, genetic damage of

Holoprosencephaly caudal dysgenesis, genetic damage of

Holoprosencephaly craniosynostosis, genetic damage of

Holoprosencephaly deletion 2p, genetic damage of

Holoprosencephaly ectrodactyly cleft lip palate, genetic damage of

Holoprosencephaly radial heart renal anomalies, genetic damage of

Holt-Oram syndrome, genetic damage of

Holzgreve Wagner Rehder syndrome, genetic damage of

Homocarnosinase deficiency, genetic damage of

Homocarnosinosis, genetic damage of

Homocystinuria, genetic damage of

Homocystinuria due to cystathionine beta-synthase, genetic damage of

Homocystinuria due to defect in methylation (cbl g), genetic damage of

Homocystinuria due to defect in methylation cbl e, genetic damage of

Homocystinuria due to defect in methylation, MTHFR deficiency, genetic damage of

Homozygous hypobetalipoproteinemia, genetic damage of

Hoon Hall syndrome, genetic damage of

Hordnes Engebretsen Knudtson syndrome, genetic damage of

Horn Kolb syndrome, genetic damage of

Horner's syndrome, genetic damage of

Hornova Dlurosova syndrome, genetic damage of

Horseshoe kidney, genetic damage of

Horton disease, genetic damage of

Horton disease, juvenile, genetic damage of

Houlston Ironton Temple syndrome, genetic damage of

Howard Young syndrome, genetic damage of

Howell-Evans syndrome, genetic damage of

Hoyeraal Hreidarsson syndrome, genetic damage of

Hoyeraal syndrome, genetic damage of

Humero spinal dysostosis congenital heart disease, genetic damage of

Humeroradial synostosis, genetic damage of

Humeroradioulnar synostosis, genetic damage of

Humerus trochlea, aplasia of, genetic damage of

Hunter Carpenter McDonald syndrome, genetic damage of

Hunter Jurenka Thompson syndrome, genetic damage of

Hunter Macpherson syndrome, genetic damage of

Hunter McAlpine syndrome, genetic damage of

Hunter McDonald syndrome, genetic damage of

Hunter Rudd Hoffmann syndrome, genetic damage of

Hunter syndrome, genetic damage of

Hunter Thomson Reed syndrome, genetic damage of

Huntington's disease, genetic damage of

Huriez scleroatrophic syndrome Hurler syndrome, genetic damage of

Hurler syndrome, genetic damage of

Hurst Hallam Hockey syndrome, genetic damage of

Hutchinson Gilford Progeria syndrome, genetic damage of

Hutchinson incisors, genetic damage of

Hutchinson-Gilford syndrome, genetic damage of

Hutteroth Spranger syndrome, genetic damage of

Hyalinosis systemic short stature, genetic damage of

Hyde Forster McCarthy Berry syndrome, genetic damage of

Hydranencephaly, genetic damage of

Hydrocephalus, genetic damage of

Hydrocephalus—Arnold Chiari—allied disorders, genetic damage of

Hydrocephalus autosomal recessive, genetic damage of

Hydrocephalus cataract microphthalmos, genetic damage of

Hydrocephalus costovertebral dysplasia Sprengel anomaly, genetic damage of

Hydrocephalus craniosynostosis bifid nose, genetic damage of

Hydrocephalus endocardial fibroelastosis cataract, genetic damage of

Hydrocephalus growth retardation skeletal anomalies, genetic damage of

Hydrocephalus obesity hypogonadism, genetic damage of

Hydrocephalus skeletal anomalies, genetic damage of

Hydrocephaly corpus callosum agenesis diaphragmatic hernia, genetic damage of

Hydrocephaly low insertion umbilicus, genetic damage of

Hydrocephaly tall stature joint laxity, genetic damage of

Hydrolethalus syndrome, genetic damage of

Hydronephrosis congenital, genetic damage of

Hydronephrosis peculiar facial expression, genetic damage of

Hydrophobia, genetic damage of (possible)

Hydrops ectrodactyly syndactyly, genetic damage of

Hydrops fetalis, genetic damage of

Hydrops fetalis anemia immune disorder absent thumb, genetic damage of

Hydroxymethylglutaricaciduria, genetic damage of

Hygroma cervical, genetic damage of

Hyper IgM syndrome, genetic damage of

Hyper-IgD syndrome, genetic damage of

Hyper-reninism, genetic damage of

Hyperadrenalism, genetic damage of

Hyperaldosteronism, genetic damage of

Hyperaldosteronism familial type 2, genetic damage of

Hyperaldosteronism, familial type 1, genetic damage of

Hyperammonemia, genetic damage of

Hyperandrogenism, genetic damage of

Hyperbilirubinemia, genetic damage of

Hyperbilirubinemia transient familial neonatal, genetic damage of

Hyperbilirubinemia type 1, genetic damage of

Hyperbilirubinemia type 2, genetic damage of

Hypercalcemia, genetic damage of

Hypercalcemia, familial benign, genetic damage of

Hypercalcemia, familial benign type 1, genetic damage of

Hypercalcemia, familial benign type 2, genetic damage of

Hypercalcemia, familial benign type 3, genetic damage of

Hypercalciuria, genetic damage of

Hypercalciuria idiopathic, genetic damage of

Hypercalciuria macular coloboma, genetic damage of

Hypercementosis, genetic damage of

Hypercholesterolemia, genetic damage of

Hypercholesterolemia due to arg3500 mutation of Apo B-100, genetic damage of

Hypercholesterolemia due to LDL receptor deficiency, genetic damage of

Hyperchylomicronemia, genetic damage of

Hypereosinophilic syndrome, genetic damage of

Hyperferritinemia, hereditary, with congenital cataracts, genetic damage of

Hyperglycerolemia, genetic damage of

Hyperglycinemia, genetic damage of

Hyperglycinemia, isolated nonketotic, genetic damage of

Hyperglycinemia, isolated nonketotic type 1, genetic damage of

Hyperglycinemia, isolated nonketotic type 2, genetic damage of

Hypergonadotropic ovarian failure, familial or sporadic, genetic damage of

Hyperhidrosis, genetic damage of

Hyperhomocysteinemia, genetic damage of

Hyperimidodipeptiduria, genetic damage of

Hyperimmunoglobinemia D with recurrent fever, genetic damage of

Hyperimmunoglobulin E—reccurrent infection syndrome, genetic damage of

Hyperimmunoglobulinemia D with periodic fever, genetic damage of

Hyperimmunoglobulinemia E, genetic damage of

Hyperinsulinism due to focal adenomatous hyperplasia, genetic damage of

Hyperinsulinism due to glucokinase deficiency, genetic damage of

Hyperinsulinism due to glutamodehydrogenase deficiency, genetic damage of

Hyperinsulinism in children, congenital, genetic damage of

Hyperinsulinism, diffuse, genetic damage of

Hyperinsulinism, focal, genetic damage of

Hyperkalemia, genetic damage of

Hyperkalemic periodic paralysis, genetic damage of

Hyperkeratosis lenticularis perstans, genetic damage of

Hyperkeratosis lenticularis perstans of Flegel, genetic damage of

Hyperkeratosis palmoplantar localized acanthokeratolytic, genetic damage of

Hyperkeratosis palmoplantar localized epidermolytic, genetic damage of

Hyperkeratosis palmoplantar with palmar crease hyperkeratosis, genetic damage of

Hyperlipoproteinemia, genetic damage of

Hyperlipoproteinemia type I, genetic damage of

Hyperlipoproteinemia type II, genetic damage of

Hyperlipoproteinemia type III, genetic damage of

Hyperlipoproteinemia type V, genetic damage of

Hyperlysinemia, genetic damage of

Hyperornithinemia, genetic damage of

Hyperornithinemia-hyperammonemia-homocitrullinuria, genetic damage of

Hyperostosid corticalis deformans juvenilis, genetic damage of

Hyperostosis cortical infantile, genetic damage of

Hyperostosis corticalis generalisata, genetic damage of

Hyperostosis frontalis interna, genetic damage of

Hyperostosis generalisata with striations, genetic damage of

Hyperoxaluria, genetic damage of

Hyperoxaluria type 1, genetic damage of

Hyperoxaluria type 2, genetic damage of

Hyperparathyroidism, genetic damage of

Hyperparathyroidism, familial, primary, genetic damage of

Hyperparathyroidism, neonatal severe primary, genetic damage of

Hyperphalangism dysmorphy bronchomalacia, genetic damage of

Hyperphenilalaninemia due to pterin-4-alpha-carbin, genetic damage of

Hyperphenylalalinemia due to dihydropteridine reductase deficiency, genetic damage of

Hyperphenylalaninemia due to 6-pyruvoyltetrahydrop, genetic damage of

Hyperphenylalaninemia due to dehydratase deficiency, genetic damage of

Hyperphenylalaninemia due to GTP cyclohydrolase deficiency, genetic damage of

Hyperphenylalaninemic embryopathy, genetic damage of

Hyperpipecolatemia, genetic damage of

Hyperprolactinemia, genetic damage of

Hyperprolinemia, genetic damage of

Hyperprolinemia type I, genetic damage of

Hyperprolinemia type II, genetic damage of

Hyperreflexia, genetic damage of

Hypersomnolence, genetic damage of

Hypertelorism and tetralogy of Fallot, genetic damage of

Hypertelorism hypospadias syndrome, genetic damage of

Hypertelorism microtia facial clefting syndrome, genetic damage of

Hypertension, genetic damage of

Hypertensive hyperkalemia, familial, genetic damage of

Hypertensive hypokalemia familial, genetic damage of

Hypertensive retinopathy, genetic damage of

Hyperthermia, genetic damage of

Hyperthermia induced defects, genetic damage of

Hyperthermia of anesthesia, genetic damage of

Hyperthyroidism due to mutations in TSH receptor, genetic damage of

Hypertrichosis atrophic skin ectropion macrostomia, genetic damage of

Hypertrichosis brachydactyly obesity and mental retardation, genetic damage of

Hypertrichosis congenital generalized X linked, genetic damage of

Hypertrichosis lanuginosa, genetic damage of

Hypertrichosis retinopathy dysmorphism, genetic damage of

Hypertrichosis universalis congenita Ambras type, genetic damage of

Hypertrichotic osteochondrodysplasia, genetic damage of

Hypertrophic branchial myopathy, genetic damage of

Hypertrophic cardiomyopathy, genetic damage of

Hypertrophic hemangiectasia, genetic damage of

Hypertrophic myocardiopathy, genetic damage of

Hypertrophic osteoarthropathy, primary or idiopathic, genetic damage of

Hypertropic neuropathy of Dejerine-Sottas, genetic damage of

Hypertryptophanemia, genetic damage of

Hypo-alphalipoproteinemia primary, genetic damage of

Hypoadrenalism, genetic damage of

Hypoadrenocorticism hypoparathyroidism moniliasis, genetic damage of

Hypoaldosteronism, genetic damage of, genetic damage of

Hypobetalipoproteinaemia ataxia hearing loss, genetic damage of

Hypobetalipoprotéinemia, familial, genetic damage of

Hypocalcemia, genetic damage of

Hypocalcemia, autosomal dominant, genetic damage of

Hypocalciuric hypercalcemia, familial, genetic damage of

Hypocalciuric hypercalcemia, familial type 1, genetic damage of

Hypocalciuric hypercalcemia, familial type 2, genetic damage of

Hypocalciuric hypercalcemia, familial type 3, genetic damage of

Hypochondrogenesis, genetic damage of

Hypochondroplasia, genetic damage of

Hypocomplementemic urticarial vasculitis, genetic damage of

Hypodermyasis, genetic damage of

Hypodontia dysplasia of nails, genetic damage of

Hypodontia of incisors and premolars, genetic damage of

Hypofibrinogenemia, familial, genetic damage of

Hypoglycemia with deficiency of glycogen synthetase in the liver, genetic damage of

Hypogonadism, genetic damage of

Hypogonadism cardiomyopathy, genetic damage of

Hypogonadism hypogonadotropic due to mutations in GR hormone, genetic damage of

Hypogonadism male mental retardation skeletal anomaly, genetic damage of

Hypogonadism mitral valve prolapse mental retardation, genetic damage of

Hypogonadism primary partial alopecia, genetic damage of

Hypogonadism retinitis pigmentosa, genetic damage of

Hypogonadism, isolated, hypogonadotropic, genetic damage of

Hypogonadotropic hypogonadism syndactyly, genetic damage of

Hypogonadotropic hypogonadism without anosmia, X linked, genetic damage of

Hypogonadotropic hypogonadism-anosmia, genetic damage of

Hypogonadotropic hypogonadism-anosmia, X linked, genetic damage of

Hypohidrotic Ectodermal Dysplasia, genetic damage of

Hypokalemia, genetic damage of

Hypokalemic alkalosis with hypercalciuria, genetic damage of

Hypokalemic periodic paralysis, genetic damage of

Hypokaliemic periodic paralysis type 1, genetic damage of

Hypoketonemic hypoglycemia, genetic damage of

Hypokinetic dilated cardiomyopathy, familial, genetic damage of

Hypomagnesemia primary, genetic damage of

Hypomandibular faciocranial dysostosis, genetic damage of

Hypomelanotic disorder, genetic damage of

Hypomelia mullerian duct anomalies, genetic damage of

Hypomentia, genetic damage of

Hypoparathyroidism, genetic damage of

Hypoparathyroidism familial isolated, genetic damage of

Hypoparathyroidism nerve deafness nephrosis, genetic damage of

Hypoparathyroidism short stature, genetic damage of

Hypoparathyroidism short stature mental retardation, genetic damage of

Hypoparathyroidism X linked, genetic damage of

Hypophosphatasia, genetic damage of

Hypophosphatasia, infantile, genetic damage of

Hypophosphatemic rickets, genetic damage of

Hypopigmentation oculocerebral syndrome Cross type, genetic damage of

Hypopituitarism, genetic damage of

Hypopituitarism micropenis cleft lip palate, genetic damage of

Hypopituitarism microphthalmia, genetic damage of

Hypopituitarism postaxial polydactyly, genetic damage of

Hypopituitary dwarfism, genetic damage of

Hypoplasia hepatic ductular, genetic damage of

Hypoplasia of the tibia with polydactyly, genetic damage of

Hypoplastic left heart syndrome, genetic damage of

Hypoplastic right heart microcephaly, genetic damage of

Hypoplastic thumb mullerian aplasia, genetic damage of

Hypoplastic thumbs hydranencephaly, genetic damage of

Hypoproconvertinemia, genetic damage of

Hypoprothrombinemia, genetic damage of

Hyporeninemic hypoaldosteronism, genetic damage of

Hyposmia nasal hypoplasia hypogonadism, genetic damage of

Hypospadias familial, genetic damage of

Hypospadias mental retardation Goldblatt type, genetic damage of

Hypotelorism cleft palate hypospadias, genetic damage of

Hypotension, orthostatic, genetic damage of (possible)

Hypothalamic dysfunction, genetic damage of

Hypothalamic hamartoblastoma syndrome, genetic damage of

Hypothalamic hamartomas, genetic damage of

Hypothyroidism, genetic damage of

Hypothyroidism due to iodide transport defect, genetic damage of

Hypothyroidism postaxial polydactyly mental retardation, genetic damage of

Hypotonic sclerotic muscular dystrophy, genetic damage of

Hypotrichosis, genetic damage of

Hypotrichosis mental retardation Lopes type, genetic damage of

Hypoxanthine guanine phosphoribosyltransferase deficiency, genetic damage of

I

I-cell disease, genetic damage of

IBIDS syndrome, genetic damage of

ICF syndrome, genetic damage of

Ichthyophobia, genetic damage of (possible)

Ichthyosiform erythroderma corneal involvement deafness, genetic damage of

Ichthyosis alopecia eclabion ectropion mental retardation, genetic damage of

Ichthyosis and male hypogonadism, genetic damage of

Ichthyosis bullosa of Siemens, genetic damage of

Ichthyosis cheek eyebrow syndrome, genetic damage of

Ichthyosis congenita, genetic damage of

Ichthyosis congenita biliary atresia, genetic damage of

Ichthyosis congenita, collodion fetus type, genetic damage of

Ichthyosis deafness mental retardation skeletal anomaly, genetic damage of

Ichthyosis exfoliativa, genetic damage of

Ichthyosis follicularis atrichia photophobia syndrome, genetic damage of

Ichthyosis hepatosplenomegaly cerebellar degeneration, genetic damage of

Ichthyosis hystrix, Curth Macklin type, genetic damage of

Ichthyosis linearis circumflexa, genetic damage of

Ichthyosis male hypogonadism, genetic damage of

Ichthyosis mental retardation Devriendt type, genetic damage of

Ichthyosis mental retardation dwarfism renal impairment, genetic damage of

Ichthyosis microphthalmos, genetic damage of

Ichthyosis tapered fingers midline groove up, genetic damage of

Ichthyosis vulgaris, genetic damage of

Ichthyosis, erythrokeratolysis hemalis, genetic damage of

Ichthyosis, keratosis follicularis spinulosa Decalvans, genetic damage of

Ichthyosis, lamellar recessive, genetic damage of

Ichthyosis, Netherton syndrome, genetic damage of

Idaho syndrome, genetic damage of

Idiopathic adolescent scoliosis, genetic damage of

Idiopathic adult neutropenia, genetic damage of

Idiopathic alveolar hypoventilation syndrome, genetic damage of

Idiopathic congenital nystagmus, dominant, X linked, genetic damage of

Idiopathic diffuse interstitial fibrosis, genetic damage of

Idiopathic dilatation of the pulmonary artery, genetic damage of

Idiopathic dilation cardiomyopathy, genetic damage of

Idiopathic double athetosis, genetic damage of

Idiopathic edema, genetic damage of

Idiopathic eosinophilic chronic pneumopathy, genetic damage of

Idiopathic facial palsy, genetic damage of

Idiopathic hypereosinophilic syndrome, genetic damage of

Idiopathic juvenile osteoporosis, genetic damage of

Idiopathic orthostatic hypotension, genetic damage of

Idiopathic pulmonary fibrosis, genetic damage of

Idiopathic pulmonary hemosiderosis, genetic damage of

Idiopathic sclerosing mesenteritis, genetic damage of

Idiopathic thrombocytopenic purpura, genetic damage of

Iduronate 2-sulfatase deficiency, genetic damage of

IFAP syndrome, genetic damage of

IgA deficiency, genetic damage of

IgA nephropathy, genetic damage of

IGDA syndrome, genetic damage of

Illum syndrome, genetic damage of

Illyngophobia, genetic damage of (possible)

Ilyina Amoashy Grygory syndrome, genetic damage of

Imaizumi Kuroki syndrome, genetic damage of

Immotile cilia syndrome, due to defective radial spokes, genetic damage of

Immotile cilia syndrome, due to excessively long cilia, genetic damage of

Immotile cilia syndrome, Kartagener type, genetic damage of

Immune deficiency, familial variable, genetic damage of

Immune thrombocytopenia, genetic damage of

Immunodeficiency microcephaly and chromosomal instability, genetic damage of

Immunodeficiency with short limb dwarfism, genetic damage of

Immunodeficiency, microcephaly with normal intelligence, genetic damage of

Imperforate anus, genetic damage of

Imperforate oropharynx costo vetebral anomalies, genetic damage of

Impossible syndrome, genetic damage of

Inborn amino acid metabolism disorder, genetic damage of

Inborn branched chain aminoaciduria, genetic damage of

Inborn error of metabolism, genetic damage of

Inborn metabolic disorder, genetic damage of

Inborn renal aminoaciduria, genetic damage of

Inborn urea cycle disorder, genetic damage of

Incisors fused, genetic damage of

Incontinentia pigmenti, genetic damage of

Incontinentia pigmenti type 1, genetic damage of

Incontinentia pigmenti type 2, genetic damage of

Infant epilepsy with migrant focal crisis, genetic damage of

Infantile apnea, genetic damage of

Infantile axonal neuropathy, genetic damage of

Infantile dysphagia, genetic damage of

Infantile multisystem inflammatory disease, genetic damage of

Infantile myofibromatosis, genetic damage of

Infantile onset spinocerebellar ataxia, genetic damage of

Infantile recurrent chronic multifocal osteomyolitis, genetic damage of

Infantile sialic acid storage disorder, genetic damage of

Infantile spasms, genetic damage of

Infantile spasms broad thumbs, genetic damage of

Infantile spinal muscular atrophy, genetic damage of

Infantile striato thalamic degeneration, genetic damage of

Inflammatory breast cancer, genetic damage of

Infundibulopelvic stenosis multicystic kidney, genetic damage of

Insensitivity to pain with anhidrosis, genetic damage of

Instability mitotic non-disjunction, genetic damage of

Insulin-resistance type B, genetic damage of

Insulinoma, genetic damage of

Intercellular cholesterol esterification disease, genetic damage of

Interferon gamma, receptor 1, deficiency, genetic damage of

Internal carotid agenesis, genetic damage of

Interstitial cystitis, genetic damage of

Interstitial pneumonia, genetic damage of

Intestinal atresia multiple, genetic damage of

Intestinal lipodystrophy, genetic damage of

Intestinal malrotation facial anomalies familial type, genetic damage of

Intestinal pseudo-obstruction chronic idiopathic, genetic damage of

Intoeing, genetic damage of

Intracranial aneurysms multiple anomaly, genetic damage of

Intracranial arterioveinous malformation, genetic damage of

Intrathoracic kidney vertebral fusion, genetic damage of

Intrauterine growth retardation mandibular malar hypoplasia, genetic damage of

Intrinsic factor, deficiency of, genetic damage of

Iophobia, genetic damage of (possible)

Iridogoniodysgenesis, dominant type, genetic damage of

Iris dysplasia hypertelorism deafness, genetic damage of

Irons Bhan syndrome, genetic damage of

Isaacs Mertens syndrome, genetic damage of

Isaacs' syndrome, genetic damage of

Ischiadic hypoplasia renal dysfunction immunodeficiency, genetic damage of

Ischiopatellar dysplasia, genetic damage of

Isochromosome 12p syndrome, genetic damage of

Isochromosome 18p, genetic damage of

Isolophobia, genetic damage of (possible)

Isotretinoin embryopathy, genetic damage of

Isthmian coarctation, genetic damage of

Ivemark syndrome, genetic damage of

Ivic syndrome, genetic damage of

J

Jackson-Weiss syndrome, genetic damage of

Jacobs syndrome, genetic damage of

Jacobsen syndrome, genetic damage of

Jadassohn Lewandowsky syndrome, genetic damage of

Jaffer Beighton syndrome, genetic damage of

Jalili syndrome, genetic damage of

Jancar syndrome, genetic damage of

Jankovic Rivera syndrome, genetic damage of

Jansen type metaphyseal chondrodysplasia, genetic damage of

Jansky-Bielschowsky disease, genetic damage of

Japanese encephalitis, genetic damage of

Jarcho-Levin syndrome, genetic damage of

Jejunal atresia, genetic damage of

Jensen syndrome, genetic damage of

Jequier Kozlowski skeletal dysplasia, genetic damage of

Jervell Lange-Nielsen syndrome, genetic damage of

Jeune syndrome, genetic damage of

Jeune syndrome situs inversus, genetic damage of

Job syndrome, genetic damage of

Johanson Blizzard syndrome, genetic damage of

Johnson Hall Krous syndrome, genetic damage of

Johnson Munson syndrome, genetic damage of

Johnston Aarons Schelley syndrome, genetic damage of

Joint instability syndrome, genetic damage of

Jones Hersh Yusk syndrome, genetic damage of

Jones syndrome, genetic damage of

Jorgenson Lenz syndrome, genetic damage of

Joseph disease, genetic damage of

Joubert syndrome, genetic damage of

Joubert syndrome bilateral chorioretinal coloboma, genetic damage of

Joubert-Boltshauser syndrome, genetic damage of

Juberg Hayward syndrome, genetic damage of

Juberg Marsidi syndrome, genetic damage of

Judge Misch Wright syndrome, genetic damage of

Jumping Frenchmen of Maine, genetic damage of

Jung Wolff Back Stahl syndrome, genetic damage of

Juvenile arthritis, genetic damage of

Juvenile cataract cerebellar atrophy myopathy mental retardation, genetic damage of

Juvenile chronic arthritis, genetic damage of

Juvenile dermatomyositis, genetic damage of

Juvenile gastrointestinal polyposis, genetic damage of

Juvenile hyaline fibromatosis, genetic damage of

Juvenile macular degeneration hypotrichosis, genetic damage of

Juvenile muscular atrophy of the distal upper limb, genetic damage of

Juvenile myoclonic epilepsy, genetic damage of

Juvenile nephronophthisis, genetic damage of

Juvenile rheumatoid arthritis, genetic damage of

Juvenile temporal arteritis, genetic damage of

K

Kabuki make up syndrome, genetic damage of

Kalam Hafeez syndrome, genetic damage of

Kaler Garrity Stern syndrome, genetic damage of

Kallikrein hypertension, genetic damage of

Kallman's syndrome, genetic damage of

Kallmann syndrome, genetic damage of

Kallmann syndrome heart disease, genetic damage of

Kallmann syndrome, type 1, X linked, genetic damage of

Kallmann syndrome, type 2, dominant, genetic damage of

Kallmann syndrome, type 3, recessive, genetic damage of

Kalyanraman syndrome, genetic damage of

Kantaputra Gorlin syndrome, genetic damage of

Kaplan Plauchu Fitch syndrome, genetic damage of

Kaplowitz Bodurtha syndrome, genetic damage of

Kaposi sarcoma, genetic damage of

Kaposiform hemangio-endothelioma, genetic damage of

Kapur Toriello syndrome, genetic damage of

Karandikar Maria Kamble syndrome, genetic damage of

Karsch Neugebauer syndrome, genetic damage of

Kartagener syndrome, genetic damage of

Kashani Strom Utley syndrome, genetic damage of

Kasznica Carlson Coppedge syndrome, genetic damage of

Katagelophobia, genetic damage of (possible)

Kathisophobia, genetic damage of (possible)

Katsantoni Papadakou Lagoyanni syndrome, genetic damage of

Katz syndrome, genetic damage of

Kaufman McKusick syndrome, genetic damage of

Kaufman oculocerebrofacial syndrome, genetic damage of

Kawasaki syndrome, genetic damage of

KBG syndrome, genetic damage of

Kearns-Sayre syndrome, genetic damage of

Keloids, genetic damage of, genetic damage of

Kennedy's disease, genetic damage of

Kennerknecht Sorgo Oberhoffer syndrome, genetic damage of

Kennerknecht Vogel syndrome, genetic damage of

Kenny Caffe syndrome, genetic damage of

Kenny syndrome, genetic damage of

Kenophobia, genetic damage of (possible)

Keratitis ichthyosis deafness syndrome, genetic damage of

Keratitis, hereditary, genetic damage of

Keratoacanthoma, genetic damage of

Keratoacanthoma familial, genetic damage of

Keratoconjunctivitis, genetic damage of

Keratoconjunctivitis sicca, genetic damage of

Keratoconus, genetic damage of

Keratoconus posticus circumscriptus, genetic damage of

Keratoderma ainhumoid and mutilans, genetic damage of

Keratoderma hypotrichosis leukonychia, genetic damage of

Keratoderma palmoplantar deafness, genetic damage of

Keratoderma palmoplantar spastic paralysis, genetic damage of

Keratoderma palmoplantaris transgrediens, genetic damage of

Keratodermia palmoplantar periorificial, genetic damage of

Keratolytic winter erythema, genetic damage of

Keratomalacia, genetic damage of

Keratosis, genetic damage of

Keratosis focal palmoplantar gingival, genetic damage of

Keratosis follicularis dwarfism cerebral atrophy, genetic damage of

Keratosis follicularis spinulosa decalvans (ichthyosis), genetic damage of

Keratosis palmoplantar-periodontopathy, genetic damage of

Keratosis palmoplantaris adenocarcinoma of the colon, genetic damage of

Keratosis palmoplantaris oesophageal colon cancer, genetic damage of

Keratosis palmoplantaris papulosa, genetic damage of

Keratosis palmoplantaris periodontopathia, genetic damage of

Keratosis palmoplantaris with corneal dystrophy, genetic damage of

Keratosis pilaris, genetic damage of

Keratosis pilaris atrophicans, genetic damage of

Keratosis, seborrheic, genetic damage of

Kernicterus, genetic damage of

Ketoaciduria mental deficiency ataxia deafness, genetic damage of

Ketotic hyperglycinemia, genetic damage of

Khalifa Graham syndrome, genetic damage of

Ki1-cell lymphoma, genetic damage of

KID syndrome, genetic damage of

Kidney and other urinary tract cancers, genetic damage of

Kienboeck disease, genetic damage of

Kikuchi disease, genetic damage of

Kimura disease, genetic damage of

Kinetophobia, genetic damage of (possible)

King-Denborough syndrome, genetic damage of

Kleeblattschaedel syndrome, genetic damage of

Klein-Waardenburg syndrome, genetic damage of

Kleine Levin syndrome, genetic damage of

Kleiner Holmes syndrome, genetic damage of

Kleptophobia, genetic damage of (possible)

Klippel Feil deformity conductive deafness absent, genetic damage of

Klippel Feil syndrome dominant type, genetic damage of

Klippel Feil syndrome recessive type, genetic damage of

Klippel Trenaunay Weber syndrome, genetic damage of

Klippel-Feil syndrome, genetic damage of

Klumpke paralysis, genetic damage of

Kluver-Bucy syndrome, genetic damage of

Kniest dysplasia, genetic damage of

Kniest like dysplasia lethal, genetic damage of

Knobloch layer syndrome, genetic damage of

Knuckle pods leuconychia sensorineural deafness, genetic damage of

Kobberling-Dunnigan syndrome, genetic damage of

Kocher-Debré-Semélaigne syndrome, genetic damage of

Kohler disease, genetic damage of

Kohlschutter Tonz syndrome, genetic damage of

Kok disease, genetic damage of

Konigsmark Knox Hussels syndrome, genetic damage of

Koone Rizzo Elias syndrome, genetic damage of

Kopophobia, genetic damage of (possible)

Korsakoff's syndrome, genetic damage of

Korula Wilson Salomon syndrome, genetic damage of

Kostmann syndrome, genetic damage of

Kosztolanyi syndrome, genetic damage of

Kotzot-Richter syndrome, genetic damage of

Koussef Nichols syndrome, genetic damage of

Kousseff syndrome, genetic damage of

Kowarski syndrome, genetic damage of

Kozlowski Brown Hardwick syndrome, genetic damage of

Kozlowski Celermajer syndrome, genetic damage of

Kozlowski Massen syndrome, genetic damage of

Kozlowski Ouvrier syndrome, genetic damage of

Kozlowski Rafinski Klicharska syndrome, genetic damage of

Kozlowski Tsuruta Taki syndrome, genetic damage of

Kozlowski Warren Fisher syndrome, genetic damage of

Kozlowski-Krajewska syndrome, genetic damage of

Krabbe leukodystrophy, genetic damage of

Krasnow Qazi syndrome, genetic damage of

Krauss Herman Holmes syndrome, genetic damage of

Krieble Bixler syndrome, genetic damage of

KTW, genetic damage of

Kufs disease, genetic damage of

Kugelberg-Welander syndrome, genetic damage of

Kumar Levick syndrome, genetic damage of

Kunze Riehm syndrome, genetic damage of

Kurczynski Casperson syndrome, genetic damage of

Kuskokwim disease, genetic damage of

Kuster Majewski Hammerstein syndrome, genetic damage of

Kuster syndrome, genetic damage of

Kuzniecky syndrome, genetic damage of

Kyasanur Forrest disease, genetic damage of

Kyphophobia, genetic damage of (possible)

Kyphosis brachyphalangy optic atrophy, genetic damage of

L

L-transposition and ccTGA, genetic damage of

Labyrinthitis syndrome, genetic damage of

Lachanophobia, genetic damage of (possible)

Lachiewicz Sibley syndrome, genetic damage of

Lacrimo-auriculo-dento-digital syndrome, genetic damage of

Lactate dehydrogenase deficiency, genetic damage of

Lactate dehydrogenase deficiency type A, genetic damage of

Lactate dehydrogenase deficiency type B, genetic damage of

Lactate dehydrogenase deficiency type C, genetic damage of

Lactic acidosis congenital infantile, genetic damage of

Lactose intolerance, genetic damage of

Ladd syndrome, genetic damage of

Ladda Zonana Ramer syndrome, genetic damage of

Lafora body disorder, genetic damage of

Lafora disease, genetic damage of

Lagophthalmia cleft lip palate, genetic damage of

Lambdoid synostosis familial, genetic damage of

Lambert syndrome, genetic damage of

Lambert-Eaton Myasthenic Syndrome (Lambert-Eaton paraneoplastic cerebellar degeneration), genetic damage of

Lambert-Eaton syndrome, genetic damage of

Lamellar ichtyosis, genetic damage of

Lamellar recessive ichthyosis, genetic damage of

Landau-Kleffner syndrome, genetic damage of

Landing disease, genetic damage of

Landouzy-Dejerine muscular dystrophy, genetic damage of

Landy Donnai syndrome, genetic damage of

Langdon Down, genetic damage of

Langer Nishino Yamaguchi syndrome, genetic damage of

Langer-Giedion syndrome, genetic damage of

Langerhans cell granulomatosis, genetic damage of

Langerhans cell histiocytosis, genetic damage of

Laparoschisis, genetic damage of

Laplane Fontaine Lagardere syndrome, genetic damage of

Large B cell diffuse lymphoma, genetic damage of

Laron syndrome, genetic damage of

Laron-type dwarfism, genetic damage of

Larsen like osseous dysplasia dwarfism, genetic damage of

Larsen like syndrome lethal type, genetic damage of

Larsen syndrome, genetic damage of

Larsen syndrome craniosynostosis, genetic damage of

Larsen syndrome, dominant type, genetic damage of

Larsen syndrome, recessive type, genetic damage of

Laryngeal abductor paralysis, genetic damage of

Laryngeal abductor paralysis mental retardation, genetic damage of

Laryngeal carcinoma, genetic damage of

Laryngeal cleft, genetic damage of

Laryngeal neoplasm, genetic damage of

Laryngeal papillomatosis, genetic damage of (possible)

Laryngeal web congenital heart disease short stature, genetic damage of

Laryngocele, genetic damage of

Laryngomalacia, genetic damage of

Laryngomalacia dominant congenital, genetic damage of

Laryngotracheoesophageal cleft pulmonary hypoplasia, genetic damage of

Larynx atresia, genetic damage of

Lassueur-Graham-Little syndrome, genetic damage of

Late onset dominant cone dystrophy, genetic damage of

Lateral body wall defect, genetic damage of

Laterality defects dominant, genetic damage of

Lattice corneal dystrophy type 2, genetic damage of

Launois-Bensaude adenolipomatosis, genetic damage of

Laurence Prosser Rocker syndrome, genetic damage of

Laurence-Moon-Bardet-Biedl syndrome, genetic damage of

Laurin Sandrow syndrome, genetic damage of

Laxova Brown Hogan syndrome, genetic damage of

LBWC—amniotic bands, genetic damage of

LBWD syndrome, genetic damage of

LCHAD deficiency, genetic damage of

LCHAD long-chain 3-hydroxyacyl-CoA dehydrogenase deficiency (G1528C mutation), genetic damage of

Leao Ribeiro Da Silva syndrome, genetic damage of

Learman syndrome, genetic damage of

Leber congenital amaurosis, genetic damage of

Leber hereditary optic neuropathy, genetic damage of

Leber miliary aneurysm, genetic damage of

Leber optic atrophy, genetic damage of

Lecithin cholesterol acyltransferase deficiency, genetic damage of

Ledderhose disease, genetic damage of

Lee Root Fenske syndrome, genetic damage of

Left ventricle-aorta tunnel, genetic damage of

Leg absence deformity cataract, genetic damage of

Legg-Calvé-Perthes syndrome, genetic damage of

Lehman syndrome, genetic damage of

Leichtman Wood Rohn syndrome, genetic damage of

Leifer Lai Buyse syndrome, genetic damage of

Leigh disease, genetic damage of

Leiner disease (complement component 5 deficiency), genetic damage of

Leiomyoma, genetic damage of

Leiomyomatosis familial, genetic damage of

Leiomyomatosis of oesophagus cataract hematuria, genetic damage of

Leiomyosarcoma, genetic damage of

Leipala Kaitila syndrome, genetic damage of

Leisti Hollister Rimoin syndrome, genetic damage of

Lemierre's syndrome, genetic damage of

Lennox-Gastaut syndrome, genetic damage of

Lentiginosis in context of NF, genetic damage of

Lenz Majewski hyperostotic dwarfism, genetic damage of

Lenz microphthalmia syndrome, genetic damage of

Leprechaunism, genetic damage of

Leprophobia, genetic damage of (possible)

Leptomeningeal capillary—venous angiomatosis, genetic damage of

Leri pleonosteosis, genetic damage of

Leri-Weil syndrome, genetic damage of

Leshima Koeda Inagaki syndrome, genetic damage of

Lethal chondrodysplasia Moerman type, genetic damage of

Lethal chondrodysplasia Seller type, genetic damage of

Lethal congenital contracture syndrome, genetic damage of

Letterer-Siwe disease, genetic damage of

Leucinosis, genetic damage of

Leukemia, genetic damage of

Leukemia subleukemic, genetic damage of

Leukemia, B-Cell, chronic, genetic damage of

Leukemia, Myeloid, genetic damage of

Leukemia, T-Cell, chronic, genetic damage of

Leukocyte adhesion deficiency syndrome, genetic damage of

Leukocyte adhesion deficiency type 2, genetic damage of

Leukocytoclastic angiitis, genetic damage of

Leukodystrophy, genetic damage of

Leukodystrophy reunion type, genetic damage of

Leukodystrophy, globoid cell, genetic damage of

Leukodystrophy, metachromatic, genetic damage of

Leukoencephalopathy palmoplantar keratoderma, genetic damage of

Leukomalacia, genetic damage of

Leukomelanoderma mental redardation hypotrichosis, genetic damage of

Leukophobia, genetic damage of (possible)

Leukoplakia, genetic damage of

Levator syndrome, genetic damage of

Levic Stefanovic Nikolic syndrome, genetic damage of

Levine Crichley syndrome, genetic damage of

Levy Hollister syndrome, genetic damage of

Lewandowski Kikolich syndrome, genetic damage of

Lewis Pashayan syndrome, genetic damage of

Lewy body dementia, genetic damage of

Lewy body disease, genetic damage of

Leydig cells hypoplasia, genetic damage of

LGCR, genetic damage of

LGS, genetic damage of

Lhermitte-Duclos disease, genetic damage of

Li-Fraumeni syndrome, genetic damage of

Lichen myxedematosus, genetic damage of

Lichen planus, genetic damage of

Lichen planus follicularis, genetic damage of

Lichen sclerosis et atrophicus, genetic damage of

Lichstenstein syndrome, genetic damage of

Lida Kannari syndrome, genetic damage of

Liddle syndrome, genetic damage of

Light chain disease, genetic damage of

Ligyrophobia, genetic damage of (possible)

Limb deficiencies distal micrognathia, genetic damage of

Limb dystonia, genetic damage of

Limb reduction defect, genetic damage of

Limb scalp and skull defects, genetic damage of

Limb transversal defect cardiac anomaly, genetic damage of

Limb-body wall complex, genetic damage of

Limb-girdle muscular dystrophy, genetic damage of

Limnophobia, genetic damage of (possible)

Lindsay Burn syndrome, genetic damage of

Lindstrom syndrome, genetic damage of

Linear hamartoma syndrome, genetic damage of

Linear nevus syndrome, genetic damage of

Linonophobia, genetic damage of (possible)

Lip lit syndrome, genetic damage of

Lipid storage myopathy, genetic damage of

Lipidosis with triglycerid storage disease, genetic damage of

Lipoamide dehydrogenase deficiency, genetic damage of

Lipoatrophic diabetes, genetic damage of

Lipodystrophy, genetic damage of

Lipodystrophy Rieger anomaly diabetes, genetic damage of

Lipogranulomatosis, genetic damage of

Lipoid congenital adrenal hyperplasia, genetic damage of

Lipoid proteinosis of Urbach and Wiethe, genetic damage of

Lipomatosis central non-encapsulated, genetic damage of

Lipomatosis familial benign cervical, genetic damage of

Lipomatosis of pancreas, congenital, genetic damage of

Lipomucopolysaccharidosis, genetic damage of

Lipoprotein disorder, genetic damage of

Liposarcoma, genetic damage of

Lisker Garcia Ramos syndrome, genetic damage of

Lison Kornbrut Feinstein syndrome, genetic damage of

Lissencephaly, genetic damage of

Lissencephaly immunodeficiency, genetic damage of

Lissencephaly syndrome type 1, genetic damage of

Lissencephaly syndrome type 2, genetic damage of

Lissencephaly, isolated, genetic damage of

Liticaphobia, genetic damage of (possible)

Liver cirrhosis, genetic damage of

Liver neoplasms, genetic damage of

Lobar atrophy of brain, genetic damage of

Lobstein disease, genetic damage of

Localized epiphyseal dysplasia, genetic damage of

Lockwood Feingold syndrome, genetic damage of

Loffredo Cennamo Cecio syndrome, genetic damage of

Logic syndrome, genetic damage of

Loiasis, genetic damage of

Loin pain hematuria syndrome, genetic damage of

Long QT Syndrome, genetic damage of

Long QT syndrome type 1, genetic damage of

Long QT syndrome type 2, genetic damage of

Long QT syndrome type 3, genetic damage of

Loose anagen hair syndrome, genetic damage of

Loose anagene syndrome, genetic damage of

Lopes Gorlin syndrome, genetic damage of

Lopes Marques de Faria syndrome, genetic damage of

Lopez Hernandez syndrome, genetic damage of

Lou Gehrig's disease, genetic damage of

Louis Bar syndrome, genetic damage of

Low birth weight dwarfism dysgammaglobulinemia, genetic damage of

Lowe Kohn Cohen syndrome, genetic damage of

Lowe oculocerebrorenal syndrome, genetic damage of

Lowe syndrome, genetic damage of

Lower limb anomaly ureteral obstruction, genetic damage of

Lower limb deficiency hypospadias, genetic damage of

Lower limb partial duplication renal agenesis, genetic damage of

Lower mesodermal defects, genetic damage of

Lowry Maclean syndrome, genetic damage of

Lowry syndrome, genetic damage of

Lowry Wood syndrome, genetic damage of

Lowry Yong syndrome, genetic damage of

LSA, genetic damage of

Lubani Al Saleh Teebi syndrome, genetic damage of

Lubinsky syndrome, genetic damage of

Lucey Driscoll syndrome, genetic damage of

Lucky Gelehrter syndrome, genetic damage of

Luiphobia, genetic damage of (possible)

Lujan-Fryns syndrome, genetic damage of

Lumbar malsegmentation short stature, genetic damage of

Lundberg syndrome, genetic damage of

Lung agenesis heart defect thumb anomalies, genetic damage of

Lung cancer, genetic damage of

Lung herniation congenital defect of sternem, genetic damage of

Lung neoplasm, genetic damage of

Lupus, genetic damage of

Lupus anticoagulant, familial, genetic damage of

Lurie Kletsky syndrome, genetic damage of

Luteinizing hormone releasing hormone, deficiency of with ataxia, genetic damage of

Lutz Richner Landolt syndrome, genetic damage of

Lutz-Lewandowsky epidermodysplasia verruciformis, genetic damage of

Lyell's syndrome, genetic damage of

Lygophobia, genetic damage of (possible)

Lymph node neoplasm, genetic damage of

Lymphadenopathy, angioimmunoblastic with dysproteinemia, genetic damage of

Lymphangiectasies lymphoedema type Hennekam type, genetic damage of

Lymphangiectasis, genetic damage of

Lymphangioleiomyomatosis, genetic damage of

Lymphangiomatosis, pulmonary, genetic damage of

Lymphangiomyomatosis, genetic damage of

Lymphatic filariasis, genetic damage of

Lymphatic neoplasm, genetic damage of

Lymphedema, genetic damage of

Lymphedema distichiasis, genetic damage of

Lymphedema hereditary type 1, genetic damage of

Lymphedema hereditary type 2, genetic damage of

Lymphedema ptosis, genetic damage of

Lymphedema, congenital, genetic damage of

Lymphoblastic lymphoma, genetic damage of

Lymphocytes; reduced or absent, genetic damage of

Lymphocytic colitis, genetic damage of

Lymphocytic infiltrate of Jessner, genetic damage of

Lymphocytic vasculitis, genetic damage of

Lymphoid hamartoma, genetic damage of

Lymphoma, genetic damage of

Lymphoma, gastric non Hodgkins type, genetic damage of

Lymphoma, large-cell, genetic damage of

Lymphoma, large-cell, immunoblastic, genetic damage of

Lymphoma, small cleaved-cell, diffuse, genetic damage of

Lymphoma, small cleaved-cell, follicular, genetic damage of

Lymphomatoid granulomatosis, genetic damage of

Lymphomatoid papulosis (LyP), genetic damage of

Lymphomatous thyroiditis, genetic damage of

Lymphosarcoma, genetic damage of

Lynch Lee Murday syndrome, genetic damage of

Lynch syndrome, genetic damage of

Lynch-Bushby syndrome, genetic damage of

Lyngstadaas syndrome, genetic damage of

LyP (lymphomatoid papulosis), genetic damage of

Lysine alpha-ketoglutarate reductase deficiency, genetic damage of

Macrophagic myofasciitis, genetic damage of

Macrosomia developmental delay dysmorphism, genetic damage of

Macrosomia microphthalmia cleft palate, genetic damage of

Macrothrombocytopenia progressive deafness, genetic damage of

Macrothrombocytopenia with leukocyte inclusions, genetic damage of

Macular corneal dystrophy, genetic damage of

Macular degeneration, genetic damage of

Macular degeneration juvenile, genetic damage of

Macular degeneration, age-related, genetic damage of

Macular degeneration, polymorphic, genetic damage of

Macular dystrophy, vitelliform, genetic damage of

Macules hereditary congenital hypopigmented and hyperpigmented, genetic damage of

Madelung's disease, genetic damage of

Madokoro Ohdo Sonoda syndrome, genetic damage of

Maffucci syndrome, genetic damage of

Mageirocophobia, genetic damage of (possible)

Maghazaji syndrome, genetic damage of

Magnesium defect in renal tubular transport of, genetic damage of

Magnesium wasting renal, genetic damage of

Mal de Debarquement, genetic damage of

Malakoplakia, genetic damage of

Male pseudohermaphroditism due to 17-beta-hydroxysteroid dehydrogenase deficiency, genetic damage of

Male pseudohermaphroditism due to 5-alpha-reductase 2 deficiency, genetic damage of

Male pseudohermaphroditism due to androgen insensitivity, genetic damage of

Male pseudohermaphroditism due to defective LH molecule, genetic damage of

Male Turner syndrome, genetic damage of

Malformations in neuronal migration, genetic damage of

Malignant astrocytoma, genetic damage of

Malignant fibrous histiocytoma, genetic damage of

Malignant germ cell tumor, genetic damage of

Malignant hyperthermia, genetic damage of

Malignant hyperthermia arthrogryposis torticollis, genetic damage of

Malignant hyperthermia susceptibility, genetic damage of

Malignant hyperthermia susceptibility type 1, genetic damage of

Malignant hyperthermia susceptibility type 2, genetic damage of

Malignant hyperthermia susceptibility type 3, genetic damage of

Malignant hyperthermia susceptibility type 4, genetic damage of

Malignant hyperthermia susceptibility type 5, genetic damage of

Malignant hyperthermia susceptibility type 6, genetic damage of

Malignant mesenchymal tumor, genetic damage of

Malignant mixed Mullerian tumor, genetic damage of

Malignant paroxysmal ventricular tachycardia, genetic damage of

Mallory-Weiss syndrome, genetic damage of

Malonic aciduria, genetic damage of

Malonyl-CoA decarboxylase deficiency, genetic damage of

Malouf syndrome, genetic damage of

Mandibuloacral dysplasia, genetic damage of

Mandibulofacial dysostosis deafness postaxial polydactly, genetic damage of

Manic depression, bipolar, genetic damage of

Manic-depressive psychosis, genetic types, genetic damage of

Mannosidosis, genetic damage of

Manouvrier syndrome, genetic damage of

Mantle cell lymphoma, genetic damage of

Maple syrup urine disease, genetic damage of

Marashi Gorlin syndrome, genetic damage of

Marchiafava Bignami disease, genetic damage of

Marchiafava-Micheli disease, genetic damage of

Marcus Gunn phenomenon, genetic damage of

Marden Walker like syndrome, genetic damage of

Marden-Walker syndrome, genetic damage of

Marek disease, genetic damage of

Marfan syndrome, genetic damage of

Marfan Syndrome type I, genetic damage of

Marfan Syndrome type II, genetic damage of

Marfan Syndrome type III, genetic damage of

Marfan Syndrome type IV, genetic damage of

Marfan Syndrome type V, genetic damage of

Marfan-Like syndrome, genetic damage of

Marfan-like syndrome, Boileau type, genetic damage of

Marfanoid build spondylolisthesis constricted pelvis, genetic damage of

Marfanoid craniosynostosis syndrome, genetic damage of

Marfanoid hypermobility, genetic damage of

Marfanoid mental retardation syndrome autosomal, genetic damage of

Marginal glioneuronal heterotopia, genetic damage of

Marie type ataxia, genetic damage of

Marie Unna congenital hypotrichosis, genetic damage of

Marineaco-Sjogren syndrome, genetic damage of

Marinesco Sjogren like syndrome, genetic damage of

Marion Mayers syndrome, genetic damage of

Markel Vikkula Mulliken syndrome, genetic damage of

Marles Greenberg Persaud syndrome, genetic damage of

Maroteaux Cohen Solal Bonaventure syndrome, genetic damage of

Maroteaux Fonfria syndrome, genetic damage of

Maroteaux Le Merrer Bensahel syndrome, genetic damage of

Maroteaux Stanescu Cousin syndrome, genetic damage of

Maroteaux Verloes Stanescu syndrome, genetic damage of

Maroteaux-Lamy syndrome, genetic damage of

Marphanoid syndrome type De Silva, genetic damage of

Marsden Nyhan Sakati syndrome, genetic damage of

Marshall syndrome, genetic damage of

Marshall-Smith syndrome, genetic damage of

Martin Bell syndrome, genetic damage of

Martinez Monasterio Pinheiro syndrome, genetic damage of

Martsolf Reed Hunter syndrome, genetic damage of

Martsolf syndrome, genetic damage of

MASA syndrome, genetic damage of

Massa Casaer Ceulemans syndrome, genetic damage of

Mast cell disease, genetic damage of

Mastigophobia, genetic damage of (possible)

Mastocytosis, genetic damage of

Mastocytosis, short stature, hearing loss, genetic damage of

Mastroiacovo De Rosa Satta syndrome, genetic damage of

Mastroiacovo Gambi Segni syndrome, genetic damage of

MAT deficiency, genetic damage of

Maternal hyperphenylalaninemia, genetic damage of

Maternally inherited diabetes and deafness, genetic damage of

Mathieu De Broca Bony syndrome, genetic damage of

Matsoukas Liarikos Giannika syndrome, genetic damage of

Matthew-Wood syndrome, genetic damage of

Maturity onset diabetes of the young, genetic damage of

Maumenee syndrome, genetic damage of

Maxillary double lip, genetic damage of

Maxillofacial dysostosis, genetic damage of

Maxillonasal dysplasia, Binder type, genetic damage of

May-Hegglin anomaly, genetic damage of

Mayer Rokitanski Kuster syndrome, genetic damage of

McAlister Crane syndrome, genetic damage of

McArdle disease, genetic damage of

McCallum Macadam Johnston syndrome, genetic damage of

McCune-Albright syndrome, genetic damage of

McDonough syndrome, genetic damage of

McDowall syndrome, genetic damage of

McGillivray syndrome, genetic damage of

McKusick Kaufman syndrome, genetic damage of

McKusick type metaphyseal chondrodysplasia, genetic damage of

McLain Debakian syndrome, genetic damage of

McPherson Clemens syndrome, genetic damage of

McPherson Robertson Cammarano syndrome, genetic damage of

Meacham Winn Culler syndrome, genetic damage of

Meadows syndrome, genetic damage of

Meckel like syndrome, genetic damage of

Meckel syndrome, genetic damage of

Medeira Dennis Donnai syndrome, genetic damage of

Median cleft lip corpus callosum lipoma skin polyps, genetic damage of

Median nodule of the upper lip, genetic damage of

Mediastinal endodermal sinus tumors, genetic damage of

Medium-chain Acyl-CoA dehydrogenase deficiency, genetic damage of

Medrano Roldan syndrome, genetic damage of

Medullary cystic disease, genetic damage of

Medullary thyroid carcinoma, genetic damage of

Medulloblastoma, genetic damage of

Megacystis microcolon intestinal hypoperistalsis syndrome, genetic damage of

Megaduodenum and/or megacystis, genetic damage of

Megaepiphyseal dwarfism, genetic damage of

Megalencephalic leukodystrophy, genetic damage of

Megalencephaly-cystic leukodystrophy, genetic damage of

Megaloblastic anemia, genetic damage of

Megalocornea mental retardation syndrome, genetic damage of

Megalocytic interstitial nephritis, genetic damage of

Mehes syndrome, genetic damage of

Mehta Lewis Patton syndrome, genetic damage of

Meier Blumberg Imahorn syndrome, genetic damage of

Meier Gorlin syndrome, genetic damage of

Meier Rotschild syndrome, genetic damage of

Meige syndrome, genetic damage of

Meigel disease, genetic damage of

Meinecke Pepper syndrome, genetic damage of

Meinecke syndrome, genetic damage of

Melanoma type 1, genetic damage of

Melanoma type 2, genetic damage of

Melanoma, familial, genetic damage of

Melanoma, malignant, genetic damage of

Melanosis neurocutaneous, genetic damage of

MELAS, genetic damage of

Meleda disease, genetic damage of

Melhem Fahl syndrome, genetic damage of

Melkersson-Rosenthal syndrome, genetic damage of

Melnick-Needles osteodysplasty, genetic damage of

Melnick-Needles syndrome, genetic damage of

Melophobia, genetic damage of (possible)

Mendelian susceptibility to atypical mycobacteria, genetic damage of

Menetrier's disease, genetic damage of

Mengel Konigsmark syndrome, genetic damage of

Meniere's disease, genetic damage of

Meningeal angiomatosis cleft hypoplastic left heart, genetic damage of

Meningioma, genetic damage of

Meningioma 1, genetic damage of

Meningocele, genetic damage of

Meningoencephalocele, genetic damage of

Meningoencephalocele-arthrogryposis-hypoplastic thumb, genetic damage of

Meningomyelocele, genetic damage of

Menkes kinky hair syndrome, genetic damage of

Menophobia, genetic damage of (possible)

Mental deficiency-epilepsy-endocrine disorders, genetic damage of

Mental mixed retardation deafnes clubbed digits, genetic damage of

Mental retardatio-polydactyly-uncombable hair, genetic damage of

Mental retardation, genetic damage of

Mental retardation anophthalmia craniosynostosis, genetic damage of

Mental retardation arachnodactyly hypotonia telangiectasia, genetic damage of

Mental retardation athetosis microphthalmia, genetic damage of

Mental retardation blepharophimosis obesity web neck, genetic damage of

Mental retardation Buenos Aires type, genetic damage of

Mental retardation cataracts calcified pinnae myopathy, genetic damage of

Mental retardation coloboma slimness, genetic damage of

Mental retardation contractural arachnodactyly, genetic damage of

Mental retardation dysmorphism hypogonadism diabetes, genetic damage of

Mental retardation epilepsy, genetic damage of

Mental retardation epilepsy bulbous nose, genetic damage of

Mental retardation gynecomastia obesity X linked, genetic damage of

Mental retardation hip luxation G6PD variant, genetic damage of

Mental retardation hypocupremia hypobetalipoproteinemia, genetic damage of

Mental retardation hypotonia skin hyperpigmentation, genetic damage of

Mental retardation macrocephaly coarse facies hypotonia, genetic damage of

Mental retardation microcephaly phalangeal facial, genetic damage of

Mental retardation microcephaly unusual facies, genetic damage of

Mental retardation Mietens Weber type, genetic damage of

Mental retardation multiple nevi, genetic damage of

Mental retardation myopathy short stature endocrine defect, genetic damage of

Mental retardation nasal hypoplasia obesity genital hypoplasia, genetic damage of

Mental retardation nasal papillomata, genetic damage of

Mental retardation osteosclerosis, genetic damage of

Mental retardation progressive spasticity, genetic damage of

Mental retardation psychosis macroorchidism, genetic damage of

Mental retardation short broad thumbs, genetic damage of

Mental retardation short stature absent phalanges, genetic damage of

Mental retardation short stature Bombay phenotype, genetic damage of

Mental retardation short stature cleft palate unusual facies, genetic damage of

Mental retardation short stature deafness genital, genetic damage of

Mental retardation short stature hand contractures genital anomalies, genetic damage of

Mental retardation short stature heart and skeletal anomalies, genetic damage of

Mental retardation short stature hypertelorism, genetic damage of

Mental retardation short stature microcephaly eye, genetic damage of

Mental retardation short stature ocular and articular anomalies, genetic damage of

Mental retardation short stature scoliosis, genetic damage of

Mental retardation short stature unusual facies, genetic damage of

Mental retardation short stature wedge shaped epiphyses, genetic damage of

Mental retardation skeletal dysplasia abducens palsy, genetic damage of

Mental retardation Smith Fineman Myers type, genetic damage of

Mental retardation spasticity ectrodactyly, genetic damage of

Mental retardation unusual facies, genetic damage of

Mental retardation unusual facies Ampola type, genetic damage of

Mental retardation unusual facies Davis Lafer type, genetic damage of

Mental retardation unusual facies talipes hand anomalies, genetic damage of

Mental retardation Wolff type, genetic damage of

Mental retardation X linked Atkin type, genetic damage of

Mental retardation X linked borderline Maoa metabolism anomaly, genetic damage of

Mental retardation X linked Brunner type, genetic damage of

Mental retardation X linked dysmorphism, genetic damage of

Mental retardation X linked dystonia dysarthria, genetic damage of

Mental retardation X linked severe Gustavson type, genetic damage of

Mental retardation X linked short stature obesity, genetic damage of

Mental retardation X linked Tranebjaerg type seizures psoriasis, genetic damage of

Mental retardation, X linked, Juberg-Marsidi type, genetic damage of

Mental retardation, X linked, Marfanoid habitus, genetic damage of

Mental retardation, X linked, nonspecific, genetic damage of

Mental retardation-unusual facies-intrauterine growth, genetic damage of

Meretoja syndrome, genetic damage of

Merkle tumors, genetic damage of

Merlob Grunebaum Reisner syndrome, genetic damage of

Merlob syndrome, genetic damage of

Mesangial sclerosis, diffuse, genetic damage of

Mesenteric panniculitis, genetic damage of

Mesodermal defects lower type, genetic damage of

Mesomelia, genetic damage of

Mesomelia radial hypoplasia bifid thumb unusual facies, genetic damage of

Mesomelia synostoses, genetic damage of

Mesomelic dwarfism cleft palate camptodactyly, genetic damage of

Mesomelic dwarfism Langer type, genetic damage of

Mesomelic dwarfism Nievergelt type, genetic damage of

Mesomelic dwarfism Reinhardt Pfeiffer type, genetic damage of

Mesomelic dysplasia skin dimples, genetic damage of

Mesomelic dysplasia Thai type, genetic damage of

Mesomelic syndrome Pfeiffer type, genetic damage of

Metabolic disorder, genetic damage of

Metacarpals 4 and 5 fusion, genetic damage of

Metachondromatosis, genetic damage of

Metachromatic leukodystrophy, genetic damage of

Metageria, genetic damage of

Metaphyseal anadysplasia, genetic damage of

Metaphyseal chondrodysplasia Schmid type, genetic damage of

Metaphyseal chondrodysplasia Spahr type, genetic damage of

Metaphyseal chondrodysplasia, McKusick type, genetic damage of

Metaphyseal chondrodysplasia, others, genetic damage of

Metaphyseal dysostosis mental retardation conductive deafness, genetic damage of

Metaphyseal dysplasia maxillary hypoplasia brachydactyly, genetic damage of

Metaphyseal dysplasia Pyle type, genetic damage of

Metastatic insulinoma, genetic damage of

Metatarsus adductus, genetic damage of

Metathesiophobia, genetic damage of (possible)

Metatrophic dysplasia, genetic damage of

Metatropic dwarfism, genetic damage of

Metatropic dysplasia 1, genetic damage of

Methylcobalamin deficiency cbl G type, genetic damage of

Methylcobalamin deficiency, cbl E complementation type, genetic damage of

Methylenetetrahydrofolate reductase deficiency, genetic damage of

Methylmalonic acidemia, genetic damage of

Methylmalonic acidemia with homocystinuria, genetic damage of

Methylmalonic aciduria microcephaly cataract, genetic damage of

Methylmalonicacidemia with homocystinuria, cbl D, genetic damage of

Methylmalonicaciduria with homocystinuria, cbl F, genetic damage of

Methylmalonicaciduria, vitamin B12 unresponsive, mut-0, genetic damage of

Methylmalonyl-Coenzyme A mutase deficiency, genetic damage of

Mevalonate kinase deficiency, genetic damage of

Mevalonicaciduria, genetic damage of

Michelin tire baby syndrome, genetic damage of

Michels Caskey syndrome, genetic damage of

Michels syndrome, genetic damage of

Mickleson syndrome, genetic damage of

Micrencephaly corpus callosum agenesis, genetic damage of

Micrencephaly olivopontocerebellar hypoplasia, genetic damage of

Micro syndrome, genetic damage of

Microbrachycephaly ptosis cleft lip, genetic damage of

Microcephalic osteodysplastic primordial dwarfism, genetic damage of

Microcephalic primordial dwarfism, genetic damage of

Microcephalic primordial dwarfism Toriello type, genetic damage of

Microcephaly, genetic damage of

Microcephaly albinism digital anomalies syndrome, genetic damage of

Microcephaly autosomal dominant, genetic damage of

Microcephaly brachydactyly kyphoscoliosis, genetic damage of

Microcephaly brain defect spasticity hypernatremia, genetic damage of

Microcephaly cardiac defect lung malsegmentation, genetic damage of

Microcephaly cardiomyopathy, genetic damage of

Microcephaly cervical spine fusion anomalies, genetic damage of

Microcephaly chorioretinopathy recessive form, genetic damage of

Microcephaly cleft palate autosomal dominant, genetic damage of

Microcephaly deafness syndrome, genetic damage of

Microcephaly developmental delay pancytopenia, genetic damage of

Microcephaly facial clefting preaxial polydactyly, genetic damage of

Microcephaly glomerulonephritis Marfanoid habitus, genetic damage of

Microcephaly hiatus hernia nephrotic syndrome, genetic damage of

Microcephaly hypergonadotropic hypogonadism short stature, genetic damage of

Microcephaly immunodeficiency and chromosomal instabilty, genetic damage of

Microcephaly immunodeficiency lymphoreticuloma, genetic damage of

Microcephaly intracranial calcification, genetic damage of

Microcephaly lymphoedema chorioretinal dysplasia, genetic damage of

Microcephaly lymphoedema syndrome, genetic damage of

Microcephaly mental retardation retinopathy, genetic damage of

Microcephaly mental retardation spasticity epilepsy, genetic damage of

Microcephaly mesobrachyphalangy tracheoesophageal fistula syndrome, genetic damage of

Microcephaly microcornea syndrome Seemanova type, genetic damage of

Microcephaly micropenis convulsions, genetic damage of

Microcephaly microphthalmos blindness, genetic damage of

Microcephaly nonsyndromal, genetic damage of

Microcephaly pontocerebellar hypoplasia dyskinesia, genetic damage of

Microcephaly seizures mental retardation heart disorders, genetic damage of

Microcephaly sparse hair mental retardation seizures, genetic damage of

Microcephaly syndactyly brachymesophalangy, genetic damage of

Microcephaly with chorioretinopathy, genetic damage of

Microcephaly with chorioretinopathy, autosomal dominant form, genetic damage of

Microcephaly with normal intelligence, immunodeficiency, genetic damage of

Microcephaly, primary autosomal recessive, genetic damage of

Microcoria, congenital, genetic damage of

Microcornea corectopia macular hypoplasia, genetic damage of

Microcornea glaucoma absent frontal sinuses, genetic damage of

Microdeletion 22 q11, genetic damage of

Microdontia hypodontia short stature, genetic damage of

Microencephaly, genetic damage of

Microgastria limb reduction defect, genetic damage of

Microgastria short stature diabetes, genetic damage of

Micromelic dwarfism Fryns type, genetic damage of

Micromelic dysplasia dislocation of radius, genetic damage of

Microphobia, genetic damage of (possible)

Microphtalmos bilateral colobomatous orbital cyst, genetic damage of

Microphthalmia, genetic damage of

Microphthalmia camptodactyly mental retardation, genetic damage of

Microphthalmia cataract, genetic damage of

Microphthalmia diaphragmatic hernia Fallot, genetic damage of

Microphthalmia mental deficiency, genetic damage of

Microphthalmia microtia fetal akinesia, genetic damage of

Microphthalmia, Lentz type, genetic damage of

Microphthalmos, microcornea, and sclerocornea, genetic damage of

Microscopic polyangiitis, genetic damage of

Microsomia hemifacial radial defects, genetic damage of

Microspherophakia metaphyseal dysplasia, genetic damage of

Microtia meatal atresia conductive deafness, genetic damage of

Microvillus inclusion disease, genetic damage of

Miculicz syndrome, genetic damage of

Midas syndrome, genetic damage of

Midline cleft of lower lip, genetic damage of

Midline defects autosomal type, genetic damage of

Midline defects recessive type, genetic damage of

Midline developmental field defects, genetic damage of

Midline field defects, genetic damage of

Midline lethal granuloma, genetic damage of

Mietens syndrome, genetic damage of

Mievis Verellen Dumoulin syndrome, genetic damage of

Mikati Najjar Sahli syndrome, genetic damage of

Mikulicz syndrome, genetic damage of

Miller Fisher syndrome, genetic damage of

Miller syndrome, genetic damage of

Miller-Dieker syndrome, genetic damage of

Milner Khallouf Gibson syndrome, genetic damage of

MILS syndrome, genetic damage of

Minkowski-Chauffard disease, genetic damage of

Miosis, congenital, genetic damage of

Mirror hands feet nasal defects, genetic damage of

Mirror polydactyly segmentation and limbs defects, genetic damage of

Misophobia, genetic damage of (possible)

Mitochondrial acetoacetyl-CoA thiolase deficiency, genetic damage of

Mitochondrial cytopathy (generic term), genetic damage of

Mitochondrial diseases of nuclear origin, genetic damage of

Mitochondrial diseases, clinically undefinite, genetic damage of

Mitochondrial encephalomyopathy aminoacidopathy, genetic damage of

Mitochondrial genetic disorders, genetic damage of

Mitochondrial myopathy lactic acidosis, genetic damage of

Mitochondrial myopathy-encephalopathy-lactic acidosis, genetic damage of

Mitochondrial PEPCK deficiency, genetic damage of

Mitochondrial trifunctional protein deficiency, genetic damage of

Mitral atresia, genetic damage of

Mitral regurgitation deafness skeletal anomalies, genetic damage of

Mitral valve prolapse, genetic damage of

Mitral valve prolapse, familial, autosomal dominant, genetic damage of

Mitral valve prolapse, familial, X linked, genetic damage of

Miura syndrome, genetic damage of

Mixed connective tissue disease, genetic damage of

Mixed Mullerian tumor, genetic damage of

Mixed sclerosing bone dystrophy, genetic damage of

MLS syndrome, genetic damage of

MMEP syndrome, genetic damage of

MMT syndrome, genetic damage of

MN1, genetic damage of

MNGIE syndrome, genetic damage of

MODY syndrome, genetic damage of

Moebius axonal neuropathy hypogonadism, genetic damage of

Moebius syndrome, genetic damage of

Moerman Vandenberghe Fryns syndrome, genetic damage of

Moeschler Clarren syndrome, genetic damage of

Mohr syndrome, genetic damage of

Molarization of anterior teeth deafness, genetic damage of

Mollica Pavone Antener syndrome, genetic damage of

Moloney syndrome, genetic damage of

Molybdenum cofactor deficiency, genetic damage of

Momo syndrome, genetic damage of

Mondini dysplasia, genetic damage of

Mondor's disease, genetic damage of

Monilethrix, genetic damage of

Monoamine oxidase A deficiency, genetic damage of

Monoclonal gammopathy of undetermined significance, genetic damage of

Monodactyly tetramelic, genetic damage of

Mononen Karnes Senac syndrome, genetic damage of

Mononeuritis multiplex, genetic damage of

Monosomy 10p, genetic damage of

Monosomy 10pter, genetic damage of

Monosomy 10q, genetic damage of

Monosomy 11 p11 p12, genetic damage of

Monosomy 11q partial, genetic damage of

Monosomy 12p12 p11, genetic damage of

Monosomy 12p13, genetic damage of

Monosomy 13q, genetic damage of

Monosomy 13q14, genetic damage of

Monosomy 13q22, genetic damage of

Monosomy 13q22, genetic damage of

Monosomy 13q32, genetic damage of

Monosomy 14q11, genetic damage of

Monosomy 14q31, genetic damage of

Monosomy 14qter, genetic damage of

Monosomy 15q1, genetic damage of

Monosomy 15q25, genetic damage of

Monosomy 17q23 q24, genetic damage of

Monosomy 18 mosaicism, genetic damage of

Monosomy 18p, genetic damage of

Monosomy 18q, genetic damage of

Monosomy 18q23, genetic damage of

Monosomy 1p, genetic damage of

Monosomy 1p22 p13, genetic damage of

Monosomy 1p31 p22, genetic damage of

Monosomy 1p32, genetic damage of

Monosomy 1p34 p32, genetic damage of

Monosomy 1q21 q25, genetic damage of

Monosomy 1q25 q32, genetic damage of

Monosomy 1q32 q42, genetic damage of

Monosomy 1q4, genetic damage of

Monosomy 20p, genetic damage of

Monosomy 21, genetic damage of

Monosomy 21q22, genetic damage of

Monosomy 2p22, genetic damage of

Monosomy 2pter p24, genetic damage of

Monosomy 2q, genetic damage of

Monosomy 2q duplication 1p, genetic damage of

Monosomy 2q24, genetic damage of

Monosomy 2q37, genetic damage of

Monosomy 3p, genetic damage of

Monosomy 3p14 p11, genetic damage of

Monosomy 3p2, genetic damage of

Monosomy 3p25, genetic damage of

Monosomy 3q13, genetic damage of

Monosomy 3q21 23, genetic damage of

Monosomy 3q27, genetic damage of

Monosomy 4p, genetic damage of

Monosomy 4p14 p16, genetic damage of

Monosomy 4q, genetic damage of

Monosomy 4q32, genetic damage of

Monosomy 5q35, genetic damage of

Monosomy 6p23, genetic damage of

Monosomy 6q, genetic damage of

Monosomy 6q1, genetic damage of

Monosomy 6q13 q15, genetic damage of

Monosomy 6q16 q21, genetic damage of

Monosomy 6q2, genetic damage of

Monosomy 7, genetic damage of

Monosomy 7q21, genetic damage of

Monosomy 7q3, genetic damage of

Monosomy 8p, genetic damage of

Monosomy 8p23 1, genetic damage of

Monosomy 8q, genetic damage of

Monosomy 8q12 21, genetic damage of

Monosomy 8q21 q22, genetic damage of

Monosomy 9p, genetic damage of

Monosomy X, genetic damage of

Monosomy Xp22 pter, genetic damage of

Monosomy Xq28, genetic damage of

Montefiore syndrome, genetic damage of

Moore Federman syndrome, genetic damage of

Moore Smith Weaver syndrome, genetic damage of

Moore Weaver syndrome, genetic damage of

Morel's ear, genetic damage of

Moreno Zachai Kaufman syndrome, genetic damage of

Morgani Turner Albright syndrome, genetic damage of

Morhosseini Holmes Walton syndrome, genetic damage of

Morillo Cucci Passarge syndrome, genetic damage of

Morphea scleroderma, genetic damage of

Morphea, generalized, genetic damage of

Morquio disease, type A, genetic damage of

Morquio disease, type B, genetic damage of

Morquio syndrome, genetic damage of

Morrison Young syndrome, genetic damage of

Morse Rawnsley Sargent syndrome, genetic damage of

Mosaic trisomy 16, genetic damage of

Mosaic variegated aneuploidy microcephaly syndrome, genetic damage of

Motor neuro-ophthalmic disorders, genetic damage of

Motor neuron disease, genetic damage of

Motor neuropathy, genetic damage of

Motor neuropathy peripheral dysautonomia, genetic damage of

Motor sensory neuropathy type 1 aplasia cutis congenita, genetic damage of

Motorphobia, genetic damage of (possible)

Mounier-Kuhn syndrome, genetic damage of

Mount Reback syndrome, genetic damage of

Mousa Al din Al Nassar syndrome, genetic damage of

Moyamoya disease, genetic damage of

Moynahan syndrome, genetic damage of

MPO deficiency, genetic damage of

MPS 1-H, genetic damage of

MPS 1-H/S, genetic damage of

MPS 1-S, genetic damage of

MPS II (mild), genetic damage of

MPS II (severe), genetic damage of

MPS III-A, genetic damage of

MPS III-B, genetic damage of

MPS III-C, genetic damage of

MPS III-D, genetic damage of

MPS IV-A, genetic damage of

Mucopolysaccharidosis type V, genetic damage of

Mucopolysaccharidosis type VI Maroteaux-Lamy—severe, intermediate, genetic damage of

Mucopolysaccharidosis type VII Sly syndrome, genetic damage of

Mucosulfatidosis, genetic damage of

Muenke syndrome, genetic damage of

Muir-Torre syndrome, genetic damage of

Mulibrey Nanism syndrome, genetic damage of

Muller Barth Menger syndrome, genetic damage of

Mullerian agenesis, genetic damage of

Mullerian aplasia, genetic damage of

Mullerian derivatives lymphangiectasia polydactyly, genetic damage of

Mullerian derivatives, persistent, genetic damage of

Mullerian duct abnormalities galactosemia, genetic damage of

Mullerian duct failure, genetic damage of

Mulliez Roux Loterman syndrome, genetic damage of

Multi-infarct dementia, genetic damage of

Multicentric osteolysis nephropathy, genetic damage of

Multicentric reticulohistiocytosis, genetic damage of

Multifocal heterotopia, genetic damage of

Multifocal motor neuropathy with conduction block, genetic damage of

Multifocal ventricular premature beats, genetic damage of

Multinodular goiter cystic kidney polydactyly, genetic damage of

Multiple acyl-CoA deficiency, genetic damage of

Multiple carboxylase deficiency, biotin responsive, genetic damage of

Multiple carboxylase deficiency, late onset, genetic damage of

Multiple carboxylase deficiency, propionic acidemia, genetic damage of

Multiple chemical sensitivity, genetic damage of

Multiple congenital anomalies mental retardation, growth failure and cleft lip palate, genetic damage of

Multiple congenital contractures, genetic damage of

Multiple contracture syndrome Finnish type, genetic damage of

Multiple endocrine neoplasia type 1, genetic damage of

Multiple endocrine neoplasia, type 2, genetic damage of

Multiple fibrofolliculoma familial, genetic damage of

Multiple hamartoma syndrome, genetic damage of

Multiple hereditary exostoses, genetic damage of

Multiple joint dislocations metaphyseal dysplasia, genetic damage of

Multiple myeloma, genetic damage of

Multiple organ failure, genetic damage of

Multiple pterygium syndrome, genetic damage of

Multiple pterygium syndrome lethal type, genetic damage of

Multiple sclerosis, genetic damage of

Multiple sclerosis ichthyosis factor VIII deficiency, genetic damage of

Multiple subcutaneous angiolipomas, genetic damage of

Multiple sulfatase deficiency, genetic damage of

Multiple synostosis syndrome, genetic damage of

Multiple system atrophy, genetic damage of

Multiple vertebral anomalies unusual facies, genetic damage of

Mulvihill Smith syndrome, genetic damage of

Munchausen by proxy syndrome, genetic damage of

MURCS association, genetic damage of

Muscle-eye-brain syndrome, genetic damage of

Muscular atrophy ataxia retinitis pigmentosa diabetes mellitus, genetic damage of

Muscular dystrophy, genetic damage of

Muscular dystrophy congenital infantile cataract hypogonadism, genetic damage of

Muscular dystrophy congenital, merosin negative, genetic damage of

Muscular dystrophy facioscapulohumeral, genetic damage of

Muscular dystrophy Hutterite type, genetic damage of

Muscular dystrophy limb girdle type 2A, Erb type, genetic damage of

Muscular dystrophy limb-girdle autosomal dominant, genetic damage of

Muscular dystrophy limb-girdle type 2B, Myoshi type, genetic damage of

Muscular dystrophy limb-girdle with beta-sarcoglycan deficiency, genetic damage of

Muscular dystrophy limb-girdle with delta-sarcoglyan deficiency, genetic damage of

Muscular dystrophy limb-girdle with gamma-sarcoglycan deficiency, genetic damage of

Muscular dystrophy white matter spongiosis, genetic damage of

Muscular dystrophy, congenital, merosin-positive, genetic damage of

Muscular dystrophy, Duchenne and Becker type, genetic damage of

Muscular fibrosis multifocal obstructed vessels, genetic damage of

Muscular phosphorylase kinase deficiency, genetic damage of

Mutations in estradiol receptor, genetic damage of

Myalgia eosinophilia associated with tryptophan, genetic damage of

Myalgic encephalomyelitis, genetic damage of

Myasthenia, familial, genetic damage of

Mycophobia, genetic damage of (possible)

Mycosis fungoides, familial, genetic damage of

Myelinopathies, genetic damage of

Myelitis, genetic damage of

Myelocerebellar disorder, genetic damage of

Myelodysplasia, genetic damage of

Myelodysplastic syndromes, genetic damage of

Myelofibrosis, genetic damage of

Myelofibrosis, idiopathic, genetic damage of

Myelofibrosis-osteosclerosis, genetic damage of

Myeloid splenomegaly, genetic damage of

Myeloperoxidase deficiency, genetic damage of

Myhre Ruvalcaba Graham syndrome, genetic damage of

Myhre Ruvalcaba Kelley syndrome, genetic damage of

Myhre School syndrome, genetic damage of

Myhre syndrome, genetic damage of

Myoadenylate deaminase deficiency, genetic damage of

Myocardium disorder, genetic damage of

Myoclonic dystonia, genetic damage of

Myoclonic progressive familial epileps, genetic damage of

Myoclonus, genetic damage of

Myoclonus ataxia, genetic damage of

Myoclonus cerebellar ataxia deafness, genetic damage of

Myoclonus epilepsy, genetic damage of

Myoclonus epilepsy partial seizure, genetic damage of

Myoclonus hereditary progressive distal muscular atrophy, genetic damage of

Myoclonus progressive epilepsy of Unverricht and Lundborg, genetic damage of

Myoclonus with epilepsy with ragged red fibers (mitochondria), genetic damage of

Myofibrillar lysis, genetic damage of

Myofibroblastic tumors, genetic damage of

Myoglobinuria, genetic damage of

Myoglobinuria dominant form, genetic damage of

Myoglobinuria recurrent, genetic damage of

Myoneurogastrointestinal encephalopathy syndrome, genetic damage of

Myopathy, genetic damage of

Myopathy and diabetes mellitus, genetic damage of

Myopathy cataract hypogonadism, genetic damage of

Myopathy congenital multicore with external ophthalmoplegia, genetic damage of

Myopathy growth and mental retardation hypospadias, genetic damage of

Myopathy Hutterite type, genetic damage of

Myopathy mitochondrial cataract, genetic damage of

Myopathy Moebius Robin syndrome, genetic damage of

Myopathy ophthalmoplegia hypoacousia areflexia, genetic damage of

Myopathy tubular agregates, genetic damage of

Myopathy with lactic acidosis and sideroblastic anemia, genetic damage of

Myopathy with lysis of myofibrils, genetic damage of

Myopathy, desmin storage, genetic damage of

Myopathy, McArdle type, genetic damage of

Myopathy, myotubular, genetic damage of

Myopathy, X linked, with excessive autophagy, genetic damage of

Myophosphorylase deficiency, genetic damage of

Myopia, infantile severe, genetic damage of

Myopia, severe, genetic damage of

Myositis, genetic damage of

Myositis ossificans, genetic damage of

Myositis ossificans post-traumatic, genetic damage of

Myositis ossificans progressiva, genetic damage of

Myositis, inclusion body, genetic damage of

Myotonia atrophica, genetic damage of

Myotonia congenita, genetic damage of

Myotonia mental retardation skeletal anomalies, genetic damage of

Myotonic dystrophy, genetic damage of

Myxedema, genetic damage of

Myxoid liposarcoma, genetic damage of

Myxoma-spotty pigmentation-endocrine overactivity, genetic damage of

N

N acetyltransferase deficiency, genetic damage of

N syndrome, genetic damage of

N-acetyl glutamate synthetase deficiency, genetic damage of

N-acetyl-alpha-D-galactosaminidase, genetic damage of

N-acetyl-alpha-glucosaminidase sulfamidase deficiency, genetic damage of

N-acetyl-glucosamine 1-phosphotransferase deficiency, genetic damage of

N-acetyl-glucosamine-6-sulfate sulfatase deficiency, genetic damage of

NADH CoQ reductase, deficiency of, genetic damage of

NADH cytochrome B5 reductase deficiency, genetic damage of

NADH diaphorase deficiency, genetic damage of

NADH methemoglobin reductase deficiency, genetic damage of

Naegelli syndrome, genetic damage of

Nager syndrome, genetic damage of

Naguib Richieri Costa syndrome, genetic damage of

Naguib syndrome, genetic damage of

Nail-patella syndrome, genetic damage of

Nakajo Nishimura syndrome, genetic damage of

Nakajo syndrome, genetic damage of

Nakamura Osame syndrome, genetic damage of

NAME syndrome, genetic damage of

Nance-Horan syndrome, genetic damage of

Nanism due to growth hormone combined deficiency, genetic damage of

Nanism due to growth hormone isolated deficiency, genetic damage of

Nanism due to growth hormone isolated deficiency with X linked hypogammaglobulinemia, genetic damage of

Nanism due to growth hormone resistance, genetic damage of

Narcolepsy, genetic damage of

Narcolepsy-cataplexy, genetic damage of

Narrow oral fissure short stature cone shaped epip, genetic damage of

Nasodigitoacoustic syndrome, genetic damage of

Nasopalpebral lipoma coloboma syndrome, genetic damage of

Nasopharyngeal carcinoma, genetic damage of

Nasopharyngeal neoplasm, genetic damage of

Nasopharyngeal teratoma Dandy Walker diaphragmatic hernia, genetic damage of

Natal teeth intestinal pseudoobstruction patent ductus, genetic damage of

Nathalie syndrome, genetic damage of

NBCC, genetic damage of

NBS, genetic damage of

Necrophobia, genetic damage of (possible)

Necrotizing encephalopathy, infantile subacute, genetic damage of

Negative rheumatoid factor polyarthritis, genetic damage of

Nelson syndrome, genetic damage of

Nemaline myopathy, genetic damage of

Neonatal hemochromatosis, genetic damage of

Neonatal ovarian cyst, genetic damage of

Neonatal transient jaundice, genetic damage of

Neopharmaphobia, genetic damage of (possible)

Neophobia, genetic damage of (possible)

Nephophobia, genetic damage of (possible)

Nephritis, IgA type, genetic damage of

Nephroblastoma, genetic damage of

Nephroblastomatosis, fetal ascites,macrosomia and Wilm's tumor, genetic damage of

Nephrocalcinosis, genetic damage of

Nephrogenic diabetes insipidus, genetic damage of

Nephrolithiasis type 2, genetic damage of

Nephronophtisis, genetic damage of

Nephronophtisis familial adult spastic quadriparesis, genetic damage of

Nephropathy deafness hyperparathyroidism, genetic damage of

Nephropathy familial with gout, genetic damage of

Nephropathy familial with hyperuricemia, genetic damage of

Nephrosclerosis, genetic damage of

Nephrosis deafness urinary tract digital malformation, genetic damage of

Nephrosis neuronal dysmigration syndrome, genetic damage of

Nephrotic syndrome ocular anomalies, genetic damage of

Nephrotic syndrome, idiopathic steroid-resistant, genetic damage of

Nerve sheath neoplasm, genetic damage of

Nesidioblastosis of pancreas, genetic damage of

Netherton syndrome ichthyosis, genetic damage of

Neu Laxova syndrome, genetic damage of

Neuhauser Daly Magnelli syndrome, genetic damage of

Neuhauser Eichner Opitz syndrome, genetic damage of

Neural crest tumor, genetic damage of

Neural tube defects X linked, genetic damage of

Neuraminidase beta-galactosidase deficiency, genetic damage of

Neuraminidase deficiency, genetic damage of

Neurasthenia, genetic damage of (possible)

Neurilemmomatosis, genetic damage of

Neuritic plaque, genetic damage of

Neuritis with brachial predilection, genetic damage of

Neuroacanthocytosis, genetic damage of

Neuroaxonal dystrophy renal tubular acidosis, genetic damage of

Neuroaxonal dystrophy, late infantile, genetic damage of

Neuroblastoma, genetic damage of

Neurocutaneous melanosis, genetic damage of

Neuroectodermal endocrine syndrome, genetic damage of

Neuroectodermal tumor, primitive, genetic damage of

Neuroectodermal tumors primitive, genetic damage of

Neuroendocrine cancer, genetic damage of

Neuroendocrine carcinoma of the cervix, genetic damage of

Neuroendocrine tumor, genetic damage of

Neuroepithelioma, genetic damage of

Neurofaciodigitorenal syndrome, genetic damage of

Neurofibrillary tangles, genetic damage of

Neurofibroma, genetic damage of

Neurofibromatosis, genetic damage of

Neurofibromatosis type 1, genetic damage of

Neurofibromatosis type 2, genetic damage of

Neurofibromatosis type 3, genetic damage of

Neurofibromatosis type 6, genetic damage of

Neurofibromatosis-Noonan syndrome, genetic damage of

Neurofibrosarcoma, genetic damage of

Neurogenic hypertension, genetic damage of

Neuroleptic malignant syndrome, genetic damage of

Neuroma biliary tract, genetic damage of

Neuronal heterotopia, genetic damage of

Neuronal interstitial dysplasia, genetic damage of

Neuronal intestinal pseudoobstruction, genetic damage of

Neuronal intranuclear hyaline inclusion disease, genetic damage of

Neuronal intranuclear inclusion disease, genetic damage of

Neuropathy ataxia and retinis pigmentosa, genetic damage of

Neuropathy congenital sensory neurotrophic keratitis, genetic damage of

Neuropathy hereditary motor and sensory lom type, genetic damage of

Neuropathy hereditary with liability to pressure palsies, genetic damage of

Neuropathy motor sensory type 2 deafness mental retardation, genetic damage of

Neuropathy sensory spastic paraplegia, genetic damage of

Neuropathy, hereditary sensory, type I, genetic damage of

Neuropathy, hereditary sensory, type II, genetic damage of

Neurotoxicity syndromes, genetic damage of

Neutral lipid storage myopathy, genetic damage of

Neutropenia and hyperlymphocytosis with large granular lymphocytes, genetic damage of

Neutropenia intermittent, genetic damage of

Neutropenia monocytopenia deafness, genetic damage of

Neutropenia, severe chronic, genetic damage of

Nevi flammei, familial multiple, genetic damage of

Nevo syndrome, genetic damage of

Nevus of ota retinitis pigmentosa, genetic damage of

Nevus sebaceus of Jadassohn, genetic damage of

Nezelof's syndrome, genetic damage of

NF-3B, genetic damage of

Nicolaides Baraitser syndrome, genetic damage of

Niemann-Pick, genetic damage of

Niemann-Pick C1 disease, genetic damage of

Niemann-Pick C2 disease, genetic damage of

Niemann-Pick disease type A, genetic damage of

Niemann-Pick Disease Type B, genetic damage of

Niemann-Pick type C, genetic damage of

Niemann-Pick type D, genetic damage of

Night blindness skeletal anomalies unusual facies, genetic damage of

Night blindness, congenital stationary, genetic damage of

Niikawa Kuroki syndrome, genetic damage of

Nivelon Nivelon Mabille syndrome, genetic damage of

Noble Bass Sherman syndrome, genetic damage of

Noctiphobia, genetic damage of (possible)

Nodular erythema digital changes, genetic damage of

Nomatophobia, genetic damage of (possible)

Non-functioning pancreatic endocrine tumor, genetic damage of

Non-Hodgkin lymphoma, genetic damage of

Non-lissencephalic cortical dysplasia, genetic damage of

Non-small cell cancer, genetic damage of

Nonallergic atopic dermatitis, genetic damage of

Noninsulin-dependent diabetes mellitus with deafness, genetic damage of

Nonketotic hyperglycinemia, genetic damage of

Nonne-Milroy disease, genetic damage of

Nonsmall cell cancer, genetic damage of

Nonsyndromal microcephaly, genetic damage of

Nonsyndromic hereditary hearing impairment, genetic damage of

Noonan like contracture myopathy hyperpyrexia, genetic damage of

Noonan like syndrome, genetic damage of

Noonan syndrome, genetic damage of

Noonanovej, genetic damage of

Norman Roberts lissencephaly syndrome, genetic damage of

Normokaliemic periodic paralysis, genetic damage of

Norrie disease, genetic damage of

Northern epilepsy, genetic damage of

Norum disease, genetic damage of

Nose polyposis, familial, genetic damage of

Nosocomephobia, genetic damage of (possible)

Nosophobia, genetic damage of (possible)

Notalgia paresthetica, genetic damage of

Nova syndrome, genetic damage of

Oculo digital syndrome, genetic damage of

Oculo facio cardio dental syndrome, genetic damage of

Oculo skeletal renal syndrome, genetic damage of

Oculo tricho anal syndrome, genetic damage of

Oculo tricho dysplasia, genetic damage of

Oculo-dento-digital syndrome, genetic damage of

Oculo-gastrointestinal muscular dystrophy, genetic damage of

Oculoauriculofrontonasal syndrome, genetic damage of

Oculoauriculovertebral dysplasia, genetic damage of

Oculocerebral hypopigmentation syndrome Cross type, genetic damage of

Oculocerebral hypopigmentation syndrome type Preus, genetic damage of

Oculocerebral syndrome with hypopigmentation, genetic damage of

Oculocerebrocutaneous syndrome, genetic damage of

Oculocerebrorenal syndrome, genetic damage of

Oculocerebrorenal syndrome of Lowe, genetic damage of

Oculocutaneous albinism, genetic damage of

Oculocutaneous albinism immunodeficiency, genetic damage of

Oculocutaneous albinism type 1, genetic damage of

Oculocutaneous albinism type 2, genetic damage of

Oculocutaneous albinism type 3, genetic damage of

Oculocutaneous albinism, tyrosinase negative, genetic damage of

Oculocutaneous albinism, tyrosinase positive, genetic damage of

Oculocutaneous tyrosinemia, genetic damage of

Oculodental syndrome Rutherfurd syndrome, genetic damage of

Oculodentodigital dysplasia dominant, genetic damage of

Oculodentodigital syndrome, genetic damage of

Oculodentoosseous dysplasia dominant, genetic damage of

Oculodentoosseous dysplasia recessive, genetic damage of

Oculomaxillofacial dysostosis, genetic damage of

Oculomelic amyoplasia, genetic damage of

Oculopalatocerebral dwarfism, genetic damage of

Oculopalatoskeletal syndrome, genetic damage of

Oculopharnygeal muscular dystrophy, genetic damage of

Oculorenocerebellar syndrome, genetic damage of

Odonto onycho dysplasia with alopecia, genetic damage of

Odontoma, genetic damage of

Odontomicronychial dysplasia, genetic damage of

Odontoonychodermal dysplasia, genetic damage of

Odontophobia, genetic damage of (possible)

Odontotrichomelic hypohidrotic dysplasia, genetic damage of

Odynophobia, genetic damage of (possible)

Oeis complex, genetic damage of

Oerter Friedman Anderson syndrome, genetic damage of

OFD syndrome type 8, genetic damage of

OFD syndrome type Figuera, genetic damage of

Ogilvie's syndrome, genetic damage of

Ohaha syndrome, genetic damage of

Ohdo Madokoro Sonoda syndrome, genetic damage of

Oikophobia, genetic damage of (possible)

Okamuto Satomura syndrome, genetic damage of

Olfactophobia, genetic damage of (possible)

Oligodactyly tetramelic postaxial, genetic damage of

Oligomeganephronic renal hypoplasia, genetic damage of

Oligomeganephrony, genetic damage of

Oligophernia, genetic damage of

Oliver McFarlane syndrome, genetic damage of

Oliver syndrome, genetic damage of

Olivopontocerebellar atrophy, genetic damage of

Olivopontocerebellar atrophy deafness, genetic damage of

Olivopontocerebellar atrophy type 1, genetic damage of

Olivopontocerebellar atrophy type 2, genetic damage of

Olivopontocerebellar atrophy type 3, genetic damage of

Ollier disease, genetic damage of

Olmsted syndrome, genetic damage of

Ombrophobia, genetic damage of (possible)

Omenn syndrome, genetic damage of

Omodysplasia, genetic damage of

Omphalocele cleft palate syndrome lethal, genetic damage of

Omphalocele exstrophy imperforate anus, genetic damage of

Omphalomesenteric cyst, genetic damage of

Onat syndrome, genetic damage of

Ondine syndrome, genetic damage of

Oneirophobia, genetic damage of (possible)

Onychonychia hypoplastic distal phalanges, genetic damage of

Onychotrichodysplasia and neutropenia, genetic damage of

Ophthalmophobia, genetic damage of (possible)

Opitz Mollica Sorge syndrome, genetic damage of

Opitz Reynolds Fitzgerald syndrome, genetic damage of

Opitz syndrome, genetic damage of

Oppositional defiant disorder, genetic damage of

Opsismodysplasia, genetic damage of

Opthalmic icthyosis, genetic damage of

Opthalmo acromelic syndrome, genetic damage of

Opthalmomandibulomelic dysplasia, genetic damage of

Opthalmoplegia ataxia hypoacusis, genetic damage of

Opthalmoplegia mental retardation lingua scrotalis, genetic damage of

Opthalmoplegia myalgia tubular aggregates, genetic damage of

Opthalmoplegia progressive external scoliosis, genetic damage of

Optic atrophy, genetic damage of

Optic atrophy opthalmoplegia ptosis deafness myopia, genetic damage of

Optic atrophy polyneuropathy deafness, genetic damage of

Optic atrophy, autosomal dominant, genetic damage of

Optic atrophy, idiopathic, autosomal recessive, genetic damage of

Optic atrophy, Leber type, genetic damage of

Optic nerve coloboma with renal disease, genetic damage of

Optic nerve disorder, genetic damage of

Optic neuritis, genetic damage of

Optic pathway glioma, genetic damage of

Opticoacoustic nerve atrophy dementia, genetic damage of

Oral facial digital syndrome, genetic damage of

Oral facial digital syndrome type 3, genetic damage of

Oral facial digital syndrome type 4, genetic damage of

Oral facial dyskinesia, genetic damage of

Oral leukoplakia, genetic damage of

Oral lichen planus, genetic damage of

Oral lichenoid lesions, genetic damage of

Oral squamous cell carcinoma, genetic damage of

Oral submucous fibrosis, genetic damage of

Oral-facial cleft, genetic damage of

Oral-facial-digital syndrome, genetic damage of

Oral-pharyngeal disorders, genetic damage of

Organic brain syndrome, genetic damage of

Organic mood syndrome, genetic damage of

Organic personality syndrome, genetic damage of

Ornithine aminotransferase deficiency, genetic damage of

Ornithine carbamoyl phosphate deficiency, genetic damage of

Ornithine carbamoyl transferase deficiency, genetic damage of

Ornithinemia, genetic damage of

Oro acral syndrome, genetic damage of

Orocraniodigital syndrome, genetic damage of

Orofaciodigital syndrome Gabrielli type, genetic damage of

Orofaciodigital syndrome Shashi type, genetic damage of

Orofaciodigital syndrome Thurston type, genetic damage of

Orofaciodigital syndrome type 2, genetic damage of

Orofaciodigital syndrome type1, genetic damage of

Orotic aciduria hereditary, genetic damage of

Orotic aciduria purines-pyrimidines, genetic damage of

Orotidylic decarboxylase deficiency, genetic damage of

Osebold Remondini syndrome, genetic damage of

Osgood-Schlatter disease, genetic damage of

Oslam syndrome, genetic damage of

Osler-Weber-Rendu syndrome, genetic damage of

Osmed syndrome, genetic damage of

Ossicular malformations, familial, genetic damage of

Osteitis deformans, genetic damage of

Osteoarthritis, genetic damage of

Osteoarthropathy of fingers familial, genetic damage of

Osteochondritis, genetic damage of

Osteochondritis deformans, genetic damage of

Osteochondritis deformans, genetic damage of

Osteochondritis deformans juvenile, genetic damage of

Osteochondritis dissecans, genetic damage of

Osteochondrodysplasia thrombocytopenia hydrocephalus, genetic damage of

Osteochondroma, genetic damage of

Osteocraniostenosis, genetic damage of

Osteodysplasia familial Anderson type, genetic damage of

Osteodysplastic dwarfism Corsello type, genetic damage of

Osteoectasia familial, genetic damage of

Osteogenesis imperfecta congenita microcephaly and cataracts, genetic damage of

Osteogenesis imperfecta congenital joint contractures, genetic damage of

Osteogenesis imperfecta retinopathy, genetic damage of

Osteogenic sarcoma, genetic damage of

Osteoglophonic dwarfism, genetic damage of

Osteolysis hereditary multicentric, genetic damage of

Osteolysis syndrome recessive, genetic damage of

Osteomalacia, genetic damage of

Osteomyelitis, genetic damage of

Osteonecrosis, genetic damage of

Osteopathia condensans disseminata with osteopoikilosis, genetic damage of

Osteopathia striata cranial sclerosis, genetic damage of

Osteopathia striata pigmentary dermopathy white forelock, genetic damage of

Osteopetrosis autosomal dominant type 1, genetic damage of

Osteopetrosis lethal, genetic damage of

Osteopetrosis renal tubular acidosis, genetic damage of

Osteopetrosis, (generic term), genetic damage of

Osteopetrosis, malignant, genetic damage of

Osteopetrosis, mild autosomal recessive form, genetic damage of

Osteopoikilosis, genetic damage of

Osteoporosis, genetic damage of

Osteoporosis macrocephaly mental retardation blindness, genetic damage of

Osteoporosis oculocutaneous hypopigmentation syndrome, genetic damage of

Osteoporosis pseudoglioma syndrome, genetic damage of

Osteosarcoma, genetic damage of

Osteosarcoma limb anomalies erythroid macrocytosis, genetic damage of

Osteosclerose type Stanescu, genetic damage of

Osteosclerosis, genetic damage of

Osteosclerosis abnormalities of nervous system and meninges, genetic damage of

Osteosclerosis autosomal dominant Worth type, genetic damage of

Ostertag type amyloidosis, genetic damage of

Ostravik Lindemann Solberg syndrome, genetic damage of

Ota Kawamura Ito syndrome, genetic damage of

Oto palato digital syndrome type I and II, genetic damage of

Oto-Palatal-digital syndrome, genetic damage of

Otodental dysplasia, genetic damage of

Otofaciocervical syndrome, genetic damage of

Otoonychoperoneal syndrome, genetic damage of

Otopalatodigital syndrome type 2, genetic damage of

Otosclerosis, genetic damage of

Otosclerosis, familial, genetic damage of

Otospondylomegaepiphyseal dysplasia; OSMED, genetic damage of

Ouvrier Billson syndrome, genetic damage of

Ovarian cancer, genetic damage of

Ovarian carcinosarcoma, genetic damage of

Ovarian dwarfism, genetic damage of

Ovarian dwarfism as part of Turner syndrome, genetic damage of

Ovarian insufficiency due to FSH resistance, genetic damage of

Ovarian remnant syndrome, genetic damage of

Overfolded helix, genetic damage of

Overgrowth radial ray defect arthrogryposis, genetic damage of

Overgrowth syndrome type Fryer, genetic damage of

Overhydrated hereditary stomatocytosis, genetic damage of

Oxalosis, genetic damage of

Oxoglutaricaciduria, genetic damage of

P

Pachydermoperiostosis, genetic damage of

Pachygyria, genetic damage of

Pachyonychia congenita Jackson Lawler type, genetic damage of

Pacman syndrome, genetic damage of

Paes Whelan Modi syndrome, genetic damage of

Paget disease extramammary, genetic damage of

Paget disease juvenile type, genetic damage of

Paget's disease, genetic damage of

Paget's disease of the bone, genetic damage of

Paget's disease of the breast, genetic damage of

Pagon Bird Detter syndrome, genetic damage of

Pagon Stephan syndrome, genetic damage of

Pai Levkoff syndrome, genetic damage of

Pai syndrome, genetic damage of

Palant cleft palate syndrome, genetic damage of

Palindromic rheumatism, genetic damage of

Pallister Mosaic syndrome, genetic damage of

Pallister-Hall syndrome, genetic damage of

Palmer Pagon syndrome, genetic damage of

Palmitoyl-protein thioesterase deficiency, genetic damage of

Palmoplantar keratoderma, genetic damage of

Palmoplantar porokeratosis of Mantoux, genetic damage of

Palsy cerebral, genetic damage of

Pancreas agenesis, genetic damage of

Pancreatic adenoma, genetic damage of

Pancreatic beta cell agenesis with neonatal diabetes mellitus, genetic damage of

Pancreatic cancer, genetic damage of

Pancreatic carcinoma, familial, genetic damage of

Pancreatic diseases, genetic damage of

Pancreatic islet cell neoplasm, genetic damage of

Pancreatic islet cell tumors, genetic damage of

Pancreatic lipomatosis duodenal stenosis, genetic damage of

Pancreatitis, hereditary, genetic damage of

Pancreatoblastoma, genetic damage of

PANDAS, genetic damage of

Panhypopituitarism, genetic damage of

Panic disorder, genetic damage of

Panmyelophthisis aplastic anemia, genetic damage of

Panniculitis, genetic damage of

Panophobia, genetic damage of (possible)

Panostotic fibrous dysplasia, genetic damage of

Panthophobia, genetic damage of (possible)

Papilledema, genetic damage of

Papillion-Lefevre syndrome, genetic damage of

Papillitis, genetic damage of

Papilloma of choroid plexus, genetic damage of

Papular mucinosis, genetic damage of

Papular urticaria, genetic damage of

Paraganglioma, genetic damage of

Paramyotonia congenita, genetic damage of

Paramyotonia congenita of Von Eulenburg, genetic damage of

Paraneoplastic cerebellar degeneration, genetic damage of

Paraomphalocele, genetic damage of

Paraparesis amyotrophy of hands and feet, genetic damage of

Paraplegia, genetic damage of

Paraplegia-brachydactyly-cone shaped epiphysis, genetic damage of

Paraplegia-mental retardation-hyperkeratosis, genetic damage of

Parapsoriasis, genetic damage of

Parastremmatic dwarfism, genetic damage of

Parathyroid cancer, genetic damage of

Parathyroid neoplasm, genetic damage of

PARC syndrome, genetic damage of

Parenchymatous cortical degeneration of cerebellum, genetic damage of

Paris-Trousseau thrombopenia, genetic damage of

Parkes-Weber syndrome, genetic damage of

Parkinson dementia Steele type, genetic damage of

Parkinson's disease, genetic damage of

Parkinsonism, genetic damage of

Parkinsonism early onset mental retardation, genetic damage of

Paroxysmal cold hemoglobinuria, genetic damage of

Paroxysmal dystonic choreoathetosis, genetic damage of

Paroxysmal nocturnal hemoglobinuria, genetic damage of

Paroxysmal ventricular fibrillation, genetic damage of

Parry-Romberg syndrome, genetic damage of

Pars planitis, genetic damage of

Parsonage Turner syndrome, genetic damage of

Partial atrioventricular canal, genetic damage of

Partial deletion of Y, genetic damage of

Partial gigantism in context of NF, genetic damage of

Partial lissencephaly, genetic damage of

Partial trisomy 8, genetic damage of

Partington Anderson syndrome, genetic damage of

Partington Mulley syndrome, genetic damage of

Parturiphobia, genetic damage of (possible)

Pascuel Castroviejo syndrome, genetic damage of

Pashayan syndrome, genetic damage of

Pat1, genetic damage of

Pat11, genetic damage of

Pat111, genetic damage of

Pat12, genetic damage of

Pat121, genetic damage of

Pat122, genetic damage of

Pat13, genetic damage of

Pat131, genetic damage of

Pat132, genetic damage of

Pat14, genetic damage of

Pat141, genetic damage of

Pat142, genetic damage of

Patau syndrome, genetic damage of

Patel Bixler syndrome, genetic damage of

Patella aplasia coxa vara tarsal synostosis, genetic damage of

Patella hypoplasia mental retardation, genetic damage of

Patent ductus arteriosus, genetic damage of

Patent ductus arteriosus familial, genetic damage of

Pathophobia, genetic damage of (possible)

Patterson Lowry syndrome, genetic damage of

Patterson pseudoleprechaunism syndrome, genetic damage of

Patterson Stevenson syndrome, genetic damage of

Pauciarticular chronic arthritis, genetic damage of

Pavone Fiumara Rizzo syndrome, genetic damage of

Pearson's marrow/pancreas syndrome, genetic damage of

Pediatric T-cell leukemia, genetic damage of

Peeling skin syndrome ichthyosis, genetic damage of

Peho syndrome, genetic damage of

Pelizaeus-Merzbacher brain sclerosis, genetic damage of

Pelizaeus-Merzbacher disease, genetic damage of

Pelizaeus-Merzbacher disease, recessive, acute infantile, genetic damage of

Pelizaeus-Merzbacher leukodystrophy, genetic damage of

Pellagra like syndrome, genetic damage of

Pellagrophobia, genetic damage of (possible)

Pelvic dysplasia arthrogryposis of lower limbs, genetic damage of

Pelvic lipomatosis, genetic damage of

Pelvic shoulder dysplasia, genetic damage of

Pemphigus, genetic damage of

Pemphigus and fogo selvagem, genetic damage of

Pemphigus foliaceus, genetic damage of

Pemphigus vulgaris, genetic damage of

Pemphigus vulgaris, familial, genetic damage of

Pena Shokeir syndrome, genetic damage of

Pendred syndrome, genetic damage of

Penis agenesia, genetic damage of

Penoscrotal transposition, genetic damage of

Penta X syndrome, genetic damage of

Pentalogy of Cantrell, genetic damage of

Pentosuria, genetic damage of

Penttinen-Aula syndrome, genetic damage of

PEPCK 1 deficiency, genetic damage of

PEPCK 2 deficiency, genetic damage of

PEPCK deficiency, mitochondrial, genetic damage of

Peptidic growth factors deficiency, genetic damage of

Periarteritis nodosa, genetic damage of

Pericardial constriction growth failure, genetic damage of

Pericardial defect diaphragmatic hernia, genetic damage of

Pericardium absent mental retardation short stature, genetic damage of

Pericardium congenital anomaly, genetic damage of

Perilymphatic fistula, genetic damage of

Perimyositis, genetic damage of

Peripartum cardiomyopathy, genetic damage of

Peripheral blood vessel disorder, genetic damage of

Peripheral nervous disorder, genetic damage of

Peripheral neuroectodermal tumor, genetic damage of

Peripheral neuropathy, genetic damage of

Peripheral T-cell lymphoma, genetic damage of

Peripheral type neurofibromatosis, genetic damage of

Perisylvian syndrome congenital bilateral, genetic damage of

Periventricular laminar heterotropia, genetic damage of

Periventricular nodular heterotopia, genetic damage of

Pernicious anemia, genetic damage of

Perniola Krajewska Carnevale syndrome, genetic damage of

Perniosis, genetic damage of

Peroxisomal bifunctional enzyme deficiency, genetic damage of

Perrault syndrome, genetic damage of

Persistent Mullerian duct syndrome, genetic damage of

Persistent truncus arteriosus, genetic damage of

Pes planus, genetic damage of

Peters anomaly, genetic damage of

Peters anomaly with short limb dwarfism, genetic damage of

Peters congenital glaucoma, genetic damage of

Peters-plus syndrome, genetic damage of

Petit Fryns syndrome, genetic damage of

Petty Laxova Wiedemann syndrome, genetic damage of

Peutz-Jeghers syndrome, genetic damage of

Peyronie disease, genetic damage of

Pfeiffer cardiocranial syndrome, genetic damage of

Pfeiffer Hirschfelder Rott syndrome, genetic damage of

Pfeiffer Kapferer syndrome, genetic damage of

Pfeiffer Mayer syndrome, genetic damage of

Pfeiffer Palm Teller syndrome, genetic damage of

Pfeiffer Rockelein syndrome, genetic damage of

Pfeiffer Singer Zschiesche syndrome, genetic damage of

Pfeiffer syndrome, genetic damage of

Pfeiffer Tietze Welte syndrome, genetic damage of

Pfeiffer type acrocephalosyndactyly, genetic damage of

Phacomatosis fourth, genetic damage of

Phacomatosis pigmentokeratotica, genetic damage of

Phacomatosis pigmentovascularis, genetic damage of

Phalacrophobia, genetic damage of (possible)

Phaoke Sharma Agarawal syndrome, genetic damage of

Pharmacophobia, genetic damage of (possible)

Phaver syndrome, genetic damage of

Phenylalanine hydroxylase deficiency, genetic damage of

Phenylalaninemia, genetic damage of

Phenylketonuria, genetic damage of

Phenylketonuria type II, genetic damage of

Phenylketonurias, genetic damage of

Phenylketonuric embryopathy, genetic damage of

Pheochromocytoma, genetic damage of

Pheochromocytoma as part of NF, genetic damage of

Philadelphia-negative chronic myeloid leukemia, genetic damage of

Phocomelia contractures absent thumb, genetic damage of

Phocomelia ectrodactyly deafness sinus arrhythmia, genetic damage of

Phocomelia Schinzel type, genetic damage of

Phocomelia syndrome, genetic damage of

Phocomelia thrombocytopenia encephalocele, genetic damage of

Phosphate diabetes, genetic damage of

Phosphoenolpyruvate carboxykinase 1 deficiency, genetic damage of

Phosphoenolpyruvate carboxykinase 2 deficiency, genetic damage of

Phosphoenolpyruvate carboxykinase deficiency, genetic damage of

Phosphoglucomutase deficiency, genetic damage of

Phosphoglucomutase deficiency type 1, genetic damage of

Phosphoglucomutase deficiency type 2, genetic damage of

Phosphoglucomutase deficiency type 3, genetic damage of

Phosphoglucomutase deficiency type 4, genetic damage of

Phosphoglycerate kinase 1 deficiency, genetic damage of

Phosphoglycerate kinase deficiency, genetic damage of

Phosphomannoisomerase deficiency, genetic damage of

Phosphoribosylpyrophosphate synthetase deficiency, genetic damage of

Photoaugliaphobia, genetic damage of (possible)

Photosensitive epilepsy, genetic damage of

Phthiriophobia, genetic damage of (possible)

Physical urticaria, genetic damage of

Phytanic acid oxidase deficiency, genetic damage of

Phytosterolemia, genetic damage of

PIBIDS syndrome, genetic damage of

Pica, genetic damage of

Picardi-Lassueur-Little syndrome, genetic damage of

Pick disease of the brain, genetic damage of

Pie Torcido, genetic damage of

Piebald trait neurologic defects, genetic damage of

Piebaldism, genetic damage of

Piepkorn Karp Hickoc syndrome, genetic damage of

Pierre Marie cerbellar ataxia, genetic damage of

Pierre Robin sequence congenital heart defect talipes, genetic damage of

Pierre Robin sequence faciodigital anomaly, genetic damage of

Pierre Robin syndrome fetal chondrodysplasia, genetic damage of

Pierre Robin syndrome hyperphalangy clinodactyly, genetic damage of

Pierre Robin syndrome oligodactyly, genetic damage of

Pierre Robin syndrome skeletal dysplasia polydactyly, genetic damage of

Pierre Robin's sequence, genetic damage of

Pierre-Robin syndrome, genetic damage of

Pigment-dispersion syndrome, genetic damage of

Pigmentary retinopathy, genetic damage of

Pigmented villonodular synovitis, genetic damage of

Pignata guarino syndrome, genetic damage of

Pili canulati, genetic damage of

Pili multigemini, genetic damage of

Pili torti, genetic damage of

Pili torti developmental delay neurological abnormalities, genetic damage of

Pili torti nerve deafness, genetic damage of

Pili torti onychodysplasia, genetic damage of

Pillay syndrome, genetic damage of

Pilo dento ungular dysplasia microcephaly, genetic damage of

Pilotto syndrome, genetic damage of

Pinealoma, genetic damage of

Pinheiro Freire Maia Miranda syndrome, genetic damage of

Pinsky Di George Harley syndrome, genetic damage of

Pipecolic acidemia, genetic damage of

PIRA, genetic damage of

Pitt Hopkins syndrome, genetic damage of

Pitt-Rogers-Danks syndrome, genetic damage of

Pituitary dwarfism, genetic damage of

Pityriasis lichenoides chronica, genetic damage of

Pityriasis rubra pilaris, genetic damage of

Piussan Lenaerts Mathieu syndrome, genetic damage of

Placenta disorders, genetic damage of

Placenta neoplasm, genetic damage of

Plagiocephaly X linked mental retardation, genetic damage of

Plasmacytoma anaplastic, genetic damage of

Plasmalogenes synthesis deficiency isolated, genetic damage of

Plasminogen activitor inhibitor type 1 deficiency, congenital, genetic damage of

Plasminogen deficiency, congenital, genetic damage of

Platelet disorder, genetic damage of

Platyspondylic lethal chondrodysplasia, genetic damage of

Platyspondyly amelogenesis imperfecta, genetic damage of

Plexosarcoma, genetic damage of

Plott syndrome, genetic damage of

Plum syndrome, genetic damage of

Plummer-Vinson syndrome, genetic damage of

Podder-Tolmie syndrome, genetic damage of

POEMS syndrome, genetic damage of

Poikiloderma atrophicans-cataract, genetic damage of

Poikiloderma congenital with bullae Weary type, genetic damage of

Poikiloderma hereditary acrokeratotic Weary type, genetic damage of

Poikiloderma of Kindler, genetic damage of

Poikiloderma of Rothmund-Thomson, genetic damage of

Poikilodermatomyositis mental retardation, genetic damage of

Poikilodermia alopecia retrognathism cleft palate, genetic damage of

Pointer syndrome, genetic damage of

Poland anomaly, genetic damage of

Poland syndrome, genetic damage of

Poliosophobia, genetic damage of (possible)

Polyarteritis, genetic damage of

Polyarteritis nodosa, genetic damage of

Polyarthritis, genetic damage of

Polyarthritis, systemic, genetic damage of

Polychondritis, genetic damage of

Polycystic kidney disease, genetic damage of

Polycystic kidney disease, adult type, genetic damage of

Polycystic kidney disease, dominant type, genetic damage of

Polycystic kidney disease, infantile type, genetic damage of

Polycystic kidney disease, recessive type, genetic damage of

Polycystic kidney disease, type 1, genetic damage of

Polycystic kidney disease, type 2, genetic damage of

Polycystic kidney disease, type 3, genetic damage of

Polycystic ovarian disease, familial, genetic damage of

Polycystic ovarian syndrome, genetic damage of

Polycystic ovaries urethral sphincter dysfunction, genetic damage of

Polycythemia vera, genetic damage of

Polydactyly, genetic damage of

Polydactyly alopecia seborrheic dermatitis, genetic damage of

Polydactyly cleft lip palate psychomotor retardation, genetic damage of

Polydactyly myopia syndrome, genetic damage of

Polydactyly postaxial, genetic damage of

Polydactyly postaxial dental and vertebral, genetic damage of

Polydactyly postaxial with median cleft of upper lip, genetic damage of

Polydactyly preaxial type 1, genetic damage of

Polydactyly syndrome middle ray duplication, genetic damage of

Polydactyly visceral anomalies cleft lip palate, genetic damage of

Polyglucosan body disease, adult, genetic damage of

Polymicrogyria turricephaly hypogenitalism, genetic damage of

Polymorphic catecholergic ventricular tachycardia, genetic damage of

Polymorphic macular degeneration, genetic damage of

Polymorphous low-grade adenocarcinoma, genetic damage of

Polymyalgia rheumatica, genetic damage of

Polymyositis, genetic damage of

Polyneuritis, genetic damage of

Polyneuropathy hand defect, genetic damage of

Polyneuropathy mental retardation acromicria prema, genetic damage of

Polyostotic fibrous dysplasia, genetic damage of

Polyposis hamartomatous intestinal, genetic damage of

Polyposis skin pigmentation alopecia fingernail changes, genetic damage of

Polyposis, familial, genetic damage of (Polyposis, familial is one of a large number of polyposis syndromes)

Polysyndactyly cardiac malformation, genetic damage of

Polysyndactyly microcephaly ptosis, genetic damage of

Polysyndactyly orofacial anomalies, genetic damage of

Polysyndactyly overgrowth syndrome, genetic damage of

Polysyndactyly trigonocephaly agenesis of corpus callosum, genetic damage of

Polysyndactyly type 4, genetic damage of

Polysyndactyly type Haas, genetic damage of

Poncet-Spiegler's cylindroma, genetic damage of

Pontoneocerebellar hypoplasia, genetic damage of

Popliteal pterygium syndrome, genetic damage of

Popliteal pterygium syndrome lethal type, genetic damage of

Porencephaly, genetic damage of

Porencephaly cerebellar hypoplasia malformations, genetic damage of

Porokeratosis of Mibelli, genetic damage of

Porokeratosis plantaris palmaris et disseminata, genetic damage of

Porokeratosis punctata palmaris et plantaris, genetic damage of

Porphyria, genetic damage of

Porphyria cutanea tarda, genetic damage of

Porphyria cutanea tarda, familial type, genetic damage of

Porphyria cutanea tarda, sporadic type, genetic damage of

Porphyria, acute intermittent, genetic damage of

Porphyria, Ala-D, genetic damage of

Porphyria, congenital erythropoietic, genetic damage of

Porphyria, hereditary coproporphyria, genetic damage of

Port wine nevi mega cisterna magna hydrocephalus, genetic damage of

Portal hypertension, genetic damage of

Portal hypertension due to infrahepatic block, genetic damage of

Portal thrombosis, genetic damage of

Portal vein thrombosis, genetic damage of

Portuguese type amyloidosis, genetic damage of

Positive rheumatoid factor polyarthritis, genetic damage of

Postaxial polydactyly mental retardation, genetic damage of

Posterior tibial tendon rupture, genetic damage of

Posterior urethral valves, genetic damage of

Posterior uveitis, genetic damage of

Posterior valve urethra, genetic damage of

Postural hypotension, genetic damage of

Potassium aggravated myotonia, genetic damage of

Potophobia, genetic damage of (possible)

Potter disease type 1, genetic damage of

Potter disease, type 3, genetic damage of

Potter sequence cleft cardiopathy, genetic damage of

Potter syndrome, genetic damage of

Potter syndrome dominant type, genetic damage of

Powell Buist Stenzel syndrome, genetic damage of

Powell Chandra Saal syndrome, genetic damage of

Powell Venencie Gordon syndrome, genetic damage of

Prader-Willi syndrome, genetic damage of

Prata Liberal Goncalves syndrome, genetic damage of

Preaxial deficiency postaxial polydactyly hypospadia, genetic damage of

Preaxial polydactyly colobomata mental retardation, genetic damage of

Precocious epileptic encephalopathy, genetic damage of

Precocious myoclonic encephalopathy, genetic damage of

Precocious puberty, genetic damage of

Precocious puberty, gonadotropin-dependant, genetic damage of

Precocious puberty, male limited, genetic damage of

Preeclampsia, genetic damage of

Preeyasombat Viravithya syndrome, genetic damage of

Pregnancy toxemia/hypertension, genetic damage of

Prekallikrein deficiency, congenital, genetic damage of

Premature aging, genetic damage of

Premature aging, Okamoto type, genetic damage of

Premature atherosclerosis photomyoclonic epilepsy, genetic damage of

Premature menopause, familial, genetic damage of

Premature ovarian failure, genetic damage of

Presbycusis, genetic damage of

Prieto Badia Mulas syndrome, genetic damage of

Prieur Griscelli syndrome, genetic damage of

Primary agammaglobulinemia, genetic damage of

Primary aldosteronism, genetic damage of

Primary amenorrhea, genetic damage of

Primary biliary cirrhosis, genetic damage of

Primary ciliary dyskinesia, genetic damage of

Primary craniosynostosis, genetic damage of

Primary cutaneous amyloidosis, genetic damage of

Primary granulocytic sarcoma, genetic damage of

Primary hyperoxaluria, genetic damage of

Primary lateral sclerosis, genetic damage of

Primary malignant lymphoma, genetic damage of

Primary pulmonary hypertension, genetic damage of

Primary sclerosing cholangitis, genetic damage of

Primary tubular proximal acidosis, genetic damage of

Primerose syndrome, genetic damage of

Primordial dwarfism, genetic damage of

Primordial microcephalic dwarfism Crachami type, genetic damage of

Prinzmetal's variant angina, genetic damage of

Procarcinoma, genetic damage of

Proconvertin deficiency, congenital, genetic damage of

Progeria, genetic damage of

Progeria short stature pigmented nevi, genetic damage of

Progeria variant syndrome Ruvalcaba type, genetic damage of

Progeroid syndrome De Barsy type, genetic damage of

Progeroid syndrome Petty type, genetic damage of

Progeroid syndrome, Penttinen type, genetic damage of

Prognathism dominant, genetic damage of

Progressive acromelanosis, genetic damage of

Progressive black carbon hyperpigmentation of infancy, genetic damage of

Progressive diaphyseal dysplasia, genetic damage of

Progressive external ophthalmoplegia, genetic damage of

Progressive hearing loss stapes fixation, genetic damage of

Progressive kinking of the hair, genetic damage of

Progressive multifocal leukoencephalopathy, genetic damage of

Progressive myositis ossificans, genetic damage of

Progressive osseous heteroplasia, genetic damage of

Progressive spinal muscular atrophy, genetic damage of

Progressive supranuclear palsy, genetic damage of

Progressive supranuclear palsy atypical, genetic damage of

Progressive systemic sclerosis, genetic damage of

Prolactinoma, familial, genetic damage of

Prolerating trichilemmal cyst, genetic damage of

Prolidase deficiency, genetic damage of

Proline oxidase deficiency, genetic damage of

Prolymphocytic leukemia, genetic damage of

Properdin deficiency, genetic damage of

Propionic acidemia, genetic damage of

Propionyl-CoA carboxylase deficiency, genetic damage of

Prosencephaly cerebellar dysgenesis, genetic damage of

Prostaglandin antenatal infection, genetic damage of

Prostate cancer, familial, genetic damage of

Prostatic malacoplakia associated with prostatic abscess, genetic damage of

Prostatitis, genetic damage of

Protein C deficiency, genetic damage of

Protein R deficiency, genetic damage of

Protein S deficiency, genetic damage of

Proteus like syndrome mental retardation eye defect, genetic damage of

Proteus syndrome, genetic damage of

Prothrombin deficiency, genetic damage of

Protoporphyria, genetic damage of

Protoporphyria, erythropoietic, genetic damage of

Proud Levine Carpenter syndrome, genetic damage of

Proximal myotonic dystrophy, genetic damage of

Proximal myotonic myopathy, genetic damage of

Proximal spinal muscular atrophy, genetic damage of

Proximal tubulopathy diabetes mellitus cerebellar ataxia, genetic damage of

Prune belly syndrome, genetic damage of

Prurigo nodularis, genetic damage of

Psellismophobia, genetic damage of (possible)

Pseudo-Gaucher disease, genetic damage of

Pseudo-Pelade of Brocq, genetic damage of

Pseudo-Turner syndrome, genetic damage of

Pseudo-Zellweger syndrome, genetic damage of

Pseudoachondroplasia, genetic damage of

Pseudoachondroplastic dysplasia, genetic damage of

Pseudoachondroplastic dysplasia 1, genetic damage of

Pseudoadrenoleukodystrophy, genetic damage of

Pseudoaminopterin syndrome, genetic damage of

Pseudoarylsulfatase A deficiency, genetic damage of

Pseudocholinesterase deficiency, genetic damage of

Pseudogout, genetic damage of

Pseudohermaphrodism anorectal anomalies, genetic damage of

Pseudohermaphroditism, genetic damage of

Pseudohermaphroditism female skeletal anomalies, genetic damage of

Pseudohermaphroditism male with gynecomastia, genetic damage of

Pseudohermaphroditism mental retardation, genetic damage of

Pseudohypoaldosteronism, genetic damage of

Pseudohypoaldosteronism type 1, genetic damage of

Pseudohypoaldosteronism type 2, genetic damage of

Pseudohypoparathyroidism, genetic damage of

Pseudomarfanism, genetic damage of

Pseudomongolism, genetic damage of

Pseudomyxoma peritonei, genetic damage of

Pseudoobstruction idiopathic intestinal, genetic damage of

Pseudopapilledema blepharophimosis hand anomalies, genetic damage of

Pseudoprogeria syndrome, genetic damage of

Pseudotoxoplasmosis syndrome, genetic damage of

Pseudotumor cerebri, genetic damage of

Pseudovaginal perineoscrotal hypospadias, genetic damage of

Pseudoxanthoma elasticum, genetic damage of

Pseudoxanthoma elasticum, dominant form, genetic damage of

Pseudoxanthoma elasticum, recessive form, genetic damage of

Psoriasis, genetic damage of

Psoriatic arthritis, genetic damage of

Psoriatic rheumatism, genetic damage of

Psychophysiologic disorders, genetic damage of

Pterigium Colli, genetic damage of

Pteromerhanophobia, genetic damage of (possible)

Pterygia mental retardation facial dysmorphism, genetic damage of

Pterygium colli mental retardation digital anomalies, genetic damage of

Pterygium of the conjunctiva, genetic damage of

Pterygium syndrome antecubital, genetic damage of

Pterygium syndrome multiple dominant type, genetic damage of

Pterygium syndrome X linked, genetic damage of

Pterygium syndrome, multiple, genetic damage of

Ptosis coloboma mental retardation, genetic damage of

Ptosis coloboma trigonocephaly, genetic damage of

Ptosis strabismus diastasis, genetic damage of

Ptosis strabismus ectopic pupils, genetic damage of

Pulmonar arterioveinous aneurysm, genetic damage of

Pulmonary agenesis, genetic damage of

Pulmonary alveolar proteinosis, genetic damage of

Pulmonary alveolar proteinosis, congenital, genetic damage of

Pulmonary arterio-veinous fistula, genetic damage of

Pulmonary artery agenesis, genetic damage of

Pulmonary artery coming from the aorta, genetic damage of

Pulmonary artery familial dilatation, genetic damage of

Pulmonary atresia with ventricular septal defect, genetic damage of

Pulmonary blastoma, genetic damage of

Pulmonary branches stenosis, genetic damage of

Pulmonary cystic lymphangiectasis, genetic damage of

Pulmonary fibrosis/granuloma, genetic damage of

Pulmonary hypertension, genetic damage of

Pulmonary hypertension, secondary, genetic damage of

Pulmonary hypoplasia familial primary, genetic damage of

Pulmonary sequestration, genetic damage of

Pulmonary supravalvular stenosis, genetic damage of

Pulmonary surfactant protein B, deficiency of, genetic damage of

Pulmonary valve stenosis, genetic damage of

Pulmonary valves agenesis, genetic damage of

Pulmonary veins stenosis, genetic damage of

Pulmonary veno-occlusive disease, genetic damage of

Pulmonary venous return anomaly, genetic damage of

Pulmonaryatresia intact ventricular septum, genetic damage of

Pulmonic stenosis with cafe-au-lait spots, genetic damage of

Punctate acrokeratoderma freckle like pigmentation, genetic damage of

Punctate inner choroidopathy, genetic damage of

Pupaphobia, genetic damage of (possible)

Pure red cell aplasia, genetic damage of

Puretic syndrome, genetic damage of

Purine nucleoside phosphorylase deficiency, genetic damage of

Purpura, genetic damage of

Purpura, Schoenlein-Henoch, genetic damage of

Purpura, thrombotic thrombocytopenic, genetic damage of

Purtilo syndrome, genetic damage of

Pycnodysostosis, genetic damage of

Pyknoachondrogenesis, genetic damage of

Pyle disease, genetic damage of

Pyoderma gangrenosum, genetic damage of

Pyrexiophobia, genetic damage of (possible)

Pyridoxine deficit, genetic damage of

Pyrimidinemia familial, genetic damage of

Pyrophobia, genetic damage of (possible)

Pyropoikilocytosis, genetic damage of

Pyrosis, genetic damage of

Pyruvate carboxylase deficiency, genetic damage of

Pyruvate decarboxylase deficiency, genetic damage of

Pyruvate dehydrogenase deficiency, genetic damage of

Pyruvate kinase deficiency, genetic damage of

Pyruvate kinase deficiency, liver type, genetic damage of

Pyruvate kinase deficiency, muscle type, genetic damage of

Q

Qazi Markouizos syndrome, genetic damage of

Quinquaud's decalvans folliculitis, genetic damage of

R

Rabson-Mendenhall syndrome, genetic damage of

Radial defect Robin sequence, genetic damage of

Radial deficiency tibial hypoplasia, genetic damage of

Radial hypoplasia triphalangeal thumbs hypospadias, genetic damage of

Radial ray agenesis, genetic damage of

Radial ray hypoplasia choanal atresia, genetic damage of

Radiation induced angiosarcoma of the breast, genetic damage of

Radiation leukemia, genetic damage of

Radiation related neoplasm/cancer, genetic damage of

Radiation syndromes, genetic damage of (possible)

Radiation-induced cancer, genetic damage of

Radicular dentin dysplasia, genetic damage of

Radiculomegaly of canine teeth congenital cataract, genetic damage of

Radio digito facial dysplasia, genetic damage of

Radio renal syndrome, genetic damage of

Radio-ulnar synostosis, genetic damage of

Radiophobia, genetic damage of (possible)

Radioulnar synostosis mental retardation hypotonia, genetic damage of

Radioulnar synostosis retinal pigment abnormalities, genetic damage of

Radius absent anogenital anomalies, genetic damage of

Raine syndrome, genetic damage of

Rambam Hasharon syndrome, genetic damage of

Rambaud Galian syndrome, genetic damage of

Ramer Ladda syndrome, genetic damage of

Ramon syndrome, genetic damage of

Ramos Arroyo Clark syndrome, genetic damage of

Ramsay Hunt paralysis syndrome, genetic damage of

Rapadilino syndrome, genetic damage of

Rapp-Hodgkin syndrome, genetic damage of

Rasmussen encephalitis, genetic damage of

Rasmussen Johnsen Thomsen syndrome, genetic damage of

Ray Peterson Scott syndrome, genetic damage of

Raynaud's disease/phenomenon, genetic damage of

Rayner Lampert Rennert syndrome, genetic damage of

Reactive arthritis, genetic damage of (possible)

Reactive attachment disorder of early childhood, genetic damage of (possible)

Reactive attachment disorder of infancy, genetic damage of (possible)

Reactive hypoglycemia, genetic damage of (possible)

Reardon Hall Slaney syndrome, genetic damage of

Reardon Wilson Cavanagh syndrome, genetic damage of

Rectal neoplasm, genetic damage of

Rectophobia, genetic damage of (possible)

Rectosigmoid neoplasm, genetic damage of

Recurrent laryngeal papillomas, genetic damage of

Recurrent peripheral facial palsy, genetic damage of

Reductional transverse limb defects, genetic damage of

Reflex sympathetic dystrophy syndrome, genetic damage of

Reflux esophagitis, genetic damage of

Refractory anemia, genetic damage of

Refsum disease, infantile form, genetic damage of

Refsum syndrome, genetic damage of

Reginato Shiapachasse syndrome, genetic damage of

Regional enteritis, genetic damage of

Reifenstein syndrome, genetic damage of

Reinhardt Pfeiffer syndrome, genetic damage of

Reiter's syndrome, genetic damage of

Renal adysplasia dominant type, genetic damage of

Renal agenesis, genetic damage of

Renal agenesis meningomyelocele mullerian defect, genetic damage of

Renal agenesis, bilateral, genetic damage of

Renal artery stenosis, genetic damage of

Renal calculi, genetic damage of

Renal caliceal diverticuli deafness, genetic damage of

Renal cancer, genetic damage of

Renal carcinoma, familial, genetic damage of

Renal cell carcinoma, genetic damage of

Renal dysplasia diffuse autosomal recessive, genetic damage of

Renal dysplasia diffuse cystic, genetic damage of

Renal dysplasia hepatic fibrosis Dandy Walker cyst, genetic damage of

Renal dysplasia limb defects, genetic damage of

Renal dysplasia megalocystis sirenomelia, genetic damage of

Renal dysplasia mesomelia radiohumeral fusion, genetic damage of

Renal failure, genetic damage of

Renal genital middle ear anomalies, genetic damage of

Renal glycosuria, genetic damage of

Renal hepatic pancreatic dysplasia Dandy Walker cyst, genetic damage of

Renal hypertension, genetic damage of

Renal osteodystrophy, genetic damage of

Renal tubular acidosis, genetic damage of

Renal tubular acidosis progressive nerve deafness, genetic damage of

Renal tubular acidosis, distal, genetic damage of

Renal tubular acidosis, distal, autosomal dominant, genetic damage of

Renal tubular acidosis, distal, autosomal recessive, genetic damage of

Renal tubular acidosis, distal, type 3, genetic damage of

Renal tubular acidosis, distal, type 4, genetic damage of

Renal tubular transport disorders, genetic damage of

Rendu-Osler-Weber disease, genetic damage of

Renier Gabreels Jasper syndrome, genetic damage of

Renoanogenital syndrome, genetic damage of

Renoprival hypertension, genetic damage of

Resistance to LH (luteinizing hormone), genetic damage of

Resistance to thyroid stimulating hormone, genetic damage of

Respiratory chain deficiency malformations, genetic damage of

Respiratory distress syndrome, adult, genetic damage of

Respiratory distress syndrome, infant, genetic damage of

Restless legs syndrome, genetic damage of

Reticulosis, familial histiocytic, genetic damage of

Retina disorder, genetic damage of

Retinal degeneration, genetic damage of

Retinal dysplasia X linked, genetic damage of

Retinal telangiectasia hypogammaglobulinemia, genetic damage of

Retinis pigmentosa deafness hypogenitalism, genetic damage of

Retinitis pigmentosa, genetic damage of

Retinitis pigmentosa mental retardation deafness, genetic damage of

Retinitis pigmentosa-deafness, genetic damage of

Retinoblastoma, genetic damage of

Retinohepatoendocrinologic syndrome, genetic damage of

Retinopathy anemia CNS anomalies, genetic damage of

Retinopathy aplastic anemia neurological abnormalities, genetic damage of

Retinopathy pigmentary mental retardation, genetic damage of

Retinopathy, arteriosclerotic, genetic damage of

Retinopathy, diabetic, genetic damage of

Retinoschisis, genetic damage of

Retinoschisis, juvenile, genetic damage of

Retinoschisis, X linked, genetic damage of

Retraction syndrome, genetic damage of

Retrolental fibroplasia, genetic damage of

Retroperitoneal fibrosis, genetic damage of

Rett like syndrome, genetic damage of

Rett syndrome, genetic damage of

Revesz Debuse syndrome, genetic damage of

Reye syndrome, genetic damage of

Reynolds Neri Hermann syndrome, genetic damage of

Reynolds syndrome, genetic damage of

Rh disease, genetic damage of

Rhabdoid tumor, genetic damage of

Rhabdomyomatous dysplasia cardiopathy genital anomalies, genetic damage of

Rhabdomyosarcoma, genetic damage of

Rhabdomyosarcoma 1, genetic damage of

Rhabdomyosarcoma 2, genetic damage of

Rhabdomyosarcoma, alveolar, genetic damage of

Rhabdomyosarcoma, embryonal, genetic damage of

Rheumatic Fever, genetic damage of (possible)

Rheumatism, genetic damage of

Rheumatoid arthritis, genetic damage of

Rheumatoid vasculitis, genetic damage of

Rhizomelic dysplasia type Patterson Lowry, genetic damage of

Rhizomelic pseudopolyarthritis, genetic damage of

Rhizomelic syndrome, genetic damage of

Rhumatoid purpura, genetic damage of

Rhypophobia, genetic damage of (possible)

Rhytiphobia, genetic damage of (possible)

Richieri Costa Colletto Otto syndrome, genetic damage of

Richieri Costa Da Silva syndrome, genetic damage of

Richieri Costa Gorlin syndrome, genetic damage of

Richieri Costa Guion Almeida acrofacial dysostosis, genetic damage of

Richieri Costa Guion Almeida Cohen syndrome, genetic damage of

Richieri Costa Guion Almeida dwarfism, genetic damage of

Richieri Costa Guion Almeida Rodini syndrome, genetic damage of

Richieri Costa Guion Almeida syndrome, genetic damage of

Richieri Costa Montagnoli syndrome, genetic damage of

Richieri Costa Orquizas syndrome, genetic damage of

Richieri Costa Silveira Pereira syndrome, genetic damage of

Richter syndrome, genetic damage of

Rieger syndrome, genetic damage of

Right atrium familial dilatation, genetic damage of

Right ventricle hypoplasia, genetic damage of

Rigid mask like face deafness polydactyly, genetic damage of

Rigid spine syndrome, genetic damage of

Riley-Day syndrome, genetic damage of

Ring chromosome 17, genetic damage of

Ringed hair disease, genetic damage of

Rippberger Aase syndrome, genetic damage of

Rivera Perez Salas syndrome, genetic damage of

Roberts syndrome, genetic damage of

Robin sequence oligodactyly, genetic damage of

Robinow Sorauf syndrome, genetic damage of

Robinow syndrome, genetic damage of

Robinow syndrome recessive form, genetic damage of

Robinson Miller Bensimon syndrome, genetic damage of

Roch-Leri mesosomatous lipomatosis, genetic damage of

Rod myopathy, genetic damage of

Rodini Richieri Costa syndrome, genetic damage of

Rokitansky Kuster Hauser syndrome, genetic damage of

Rokitansky sequence, genetic damage of

Romano-Ward syndrome, genetic damage of

Romberg hemi-facial atrophy, genetic damage of

Rombo syndrome, genetic damage of

Rommen Mueller Sybert syndrome, genetic damage of

Rosai-Dorfman disease, genetic damage of

Rosenberg Chutorian syndrome, genetic damage of

Rosenberg Lohr syndrome, genetic damage of

Rothmund-Thomson syndrome, genetic damage of

Rotor syndrome, genetic damage of

Roussy Levy hereditary areflexic dystasia, genetic damage of

Roussy-Levy syndrome, genetic damage of

Roy Maroteaux Kremp syndrome, genetic damage of

Rozin Hertz Goodman syndrome, genetic damage of

Rubinstein Taybi like syndrome, genetic damage of

Rubinstein-Taybi syndrome, genetic damage of

Rudd Klimek syndrome, genetic damage of

Rudiger syndrome, genetic damage of

Rumination disorder, genetic damage of (possible)

Rupophobia, genetic damage of (possible)

Rutledge Friedman Harrod syndrome, genetic damage of

Ruvalcaba Churesigaew Myhre syndrome, genetic damage of

Ruvalcaba syndrome, genetic damage of

Ruvalcaba-Myhre syndrome, genetic damage of

Ruvalcaba-Myhre-Smith syndrome (BRR), genetic damage of

Ruzicka Goerz Anton syndrome, genetic damage of

S

Saal Bulas syndrome, genetic damage of

Saal Greenstein syndrome, genetic damage of

Sabinas brittle hair syndrome, genetic damage of

Saccharopinuria, genetic damage of

Sackey Sakati Aur syndrome, genetic damage of

Sacral agenesis, genetic damage of

Sacral defect anterior sacral meningocele, genetic damage of

Sacral hemangiomas multiple congenital abnormaliti, genetic damage of

Sacral meningocele conotruncal heart defects, genetic damage of

Sacrococcygeal dysgenesis association, genetic damage of

Saethre-Chotzen syndrome, genetic damage of

Saito Kuba Tsuruta syndrome, genetic damage of

Sakati syndrome, genetic damage of

Salcedo syndrome, genetic damage of

Saldino Mainzer syndrome, genetic damage of

Salivary disorder, genetic damage of

Salivary gland disorders, genetic damage of

Salla disease, genetic damage of

Sallis Beighton syndrome, genetic damage of

Salti Salem syndrome, genetic damage of

Sammartino Decreccio syndrome, genetic damage of

Samson Gardner syndrome, genetic damage of

Samson Viljoen syndrome, genetic damage of

Sanderson Fraser syndrome, genetic damage of

Sandhaus Ben Ami syndrome, genetic damage of

Sandhoff disease, genetic damage of

Sanfilippo syndrome, genetic damage of

Sanfilippo syndrome type A, genetic damage of

Sanfilippo syndrome type B, genetic damage of

Sanfilippo syndrome type C, genetic damage of

Sanfilippo syndrome type D, genetic damage of

Schereshevskij Turner, genetic damage of

Scheurermann's disease, genetic damage of

Schimke syndrome, genetic damage of

Schindler disease, genetic damage of

Schinzel acrocallosal syndrome, genetic damage of

Schinzel Giedion syndrome, genetic damage of

Schinzel syndrome, genetic damage of

Schinzel-Giedion midface retraction syndrome, genetic damage of

Schisis association, genetic damage of

Schizencephaly, genetic damage of

Schizophrenia mental retardation deafness retinitis, genetic damage of

Schizophrenia, genetic types, genetic damage of

Schlegelberger Grote syndrome, genetic damage of

Schmidt syndrome, genetic damage of

Schmitt Gillenwater Kelly syndrome, genetic damage of

Schneckenbecken dysplasia, genetic damage of

Schofer Beetz Bohl syndrome, genetic damage of

Scholte Begeer Van Essen syndrome, genetic damage of

Schonlein-Henoch purpura, genetic damage of

Schraderman's disease, genetic damage of

Schrander Stumpel Theunissen Hulsmans syndrome, genetic damage of

Schroer Hammer Mauldin syndrome, genetic damage of

Schwannoma, malignant, genetic damage of

Schwannomatosis, genetic damage of

Schwartz Newark syndrome, genetic damage of

Schwartz-Jampel syndrome, genetic damage of

Schweitzer Kemink Malcolm syndrome, genetic damage of

Scimitar syndrome, genetic damage of

Sciophobia, genetic damage of (possible)

Scleroatonic myopathy, genetic damage of

Sclerocornea syndactyly ambiguous genitalia, genetic damage of

Scleroderma, genetic damage of

Scleromyxedema, genetic damage of

Sclerosing bone dysplasia mental retardation, genetic damage of

Sclerosing cholangitis, genetic damage of

Sclerosing Mesenteritis, genetic damage of

Sclerosteosis, genetic damage of

Scoditti Geminiani Colonna syndrome, genetic damage of

Scoleciphobia, genetic damage of (possible)

Scoliosis as part of NF, genetic damage of

Scoliosis with unilateral unsegmented bar, genetic damage of

Scopophobia, genetic damage of (possible)

SCOT deficiency, genetic damage of

Scotomaphobia, genetic damage of (possible)

Scott Aarskog syndrome, genetic damage of

Scott Bryant Graham syndrome, genetic damage of

Scott craniodigital syndrome with mental retardation, genetic damage of

Scott syndrome, genetic damage of

Sea-blue histiocytosis, genetic damage of

Seaver Cassidy syndrome, genetic damage of

Sebocystomatosis, genetic damage of

Seborrheic keratosis, genetic damage of

Seckel like syndrome Majoor Krakauer type, genetic damage of

Seckel like syndrome type Buebel, genetic damage of

Seckel syndrome, genetic damage of

Secondary pulmonary hypertension, genetic damage of

Seemanova Lesny syndrome, genetic damage of

Seemanova syndrome type 2, genetic damage of

Segawa syndrome, genetic damage of

Seghers syndrome, genetic damage of

Segmental neurofibromatosis, genetic damage of

Segmental vertebral anomalies, genetic damage of

Seitelberger disease, genetic damage of

Seizures benign familial neonatal recessive form, genetic damage of

Seizures mental retardation hair dysplasia, genetic damage of

Selachophobia, genetic damage of (possible)

Selenophobia, genetic damage of (possible)

Selig Benacerraf Greene syndrome, genetic damage of

Seminoma, genetic damage of

Semmerkrot Haraldsson Weenaes syndrome, genetic damage of

Sengers Hamel Otten syndrome, genetic damage of

Senior syndrome, genetic damage of

Sensorineural hearing loss, genetic damage of (possible)

Sensory neuropathy, genetic damage of

Sensory neuropathy type 1, genetic damage of

Sensory radicular neuropathy recessive form, genetic damage of

Senter syndrome, genetic damage of

Seow Najjar syndrome, genetic damage of

Seplophobia, genetic damage of (possible)

Septo-optic dysplasia, genetic damage of

Septooptic dysplasia, genetic damage of

Septooptic dysplasia digital anomalies, genetic damage of

Sequeiros Sack syndrome, genetic damage of

Seres Santamaria Arimany Muniz syndrome, genetic damage of

Setleis syndrome, genetic damage of

Severe combined immunodeficiency, genetic damage of

Severe combined immunodeficiency, alymphocytotic type, genetic damage of

Severe combined immunodeficiency, HLA class 2-negative, genetic damage of

Severe infantile axonal neuropathy, genetic damage of

Sexual precocity, familial, gonadotropin-independent, genetic damage of

Sezary syndrome, genetic damage of

Sezary's lymphoma, genetic damage of

Shapiro syndrome, genetic damage of

Sharma Kapoor Ramji syndrome, genetic damage of

Sharp syndrome, genetic damage of

Sheehan syndrome, genetic damage of

Shith Filkins syndrome, genetic damage of

Shokeir syndrome, genetic damage of

Short broad great toe macrocranium, genetic damage of

Short chain Acyl CoA dehydrogenase deficiency, genetic damage of

Short limb dwarf lethal Colavita Kozlowski type, genetic damage of

Short limb dwarf lethal Mcalister Crane type, genetic damage of

Short limb dwarf mental retardation myopia, genetic damage of

Short limb dwarf oedema iris coloboma, genetic damage of

Short limb dwarfism Al Gazali type, genetic damage of

Short limbs abnormal face congenital heart disease, genetic damage of

Short limbs subluxed knees cleft palate, genetic damage of

Short rib syndrome Beemer type, genetic damage of

Short rib-polydactyly syndrome, genetic damage of

Short rib-polydactyly syndrome, Beermer type, genetic damage of

Short rib-polydactyly syndrome, Majewski type, genetic damage of

Short rib-polydactyly syndrome, Saldino-Noonan type, genetic damage of

Short rib-polydactyly syndrome, Verma-Naumoff type, genetic damage of

Short ribs craniosynostosis polysyndactyly, genetic damage of

Short stature abnormal skin pigmentation mental retardation, genetic damage of

Short stature Brussels type, genetic damage of

Short stature contractures hypotonia, genetic damage of

Short stature cranial hyperostosis hepatomegaly, genetic damage of

Short stature deafness neutrophil dysfunction, genetic damage of

Short stature dysmorphic face pelvic scapula dysplasia, genetic damage of

Short stature heart defect craniofacial anomalies, genetic damage of

Short stature hyperkaliemia acidosis, genetic damage of

Short stature locking fingers, genetic damage of

Short stature mental retardation eye anomalies, genetic damage of

Short stature mental retardation eye defects, genetic damage of

Short stature microcephaly heart defect, genetic damage of

Short stature microcephaly seizures deafness, genetic damage of

Short stature monodactylous ectrodactyly cleft palate, genetic damage of

Short stature prognathism short femoral necks, genetic damage of

Short stature Robin sequence cleft mandible hand anomalies clubfoot, genetic damage of

Short stature talipes natal teeth, genetic damage of

Short stature valvular heart disease, genetic damage of

Short stature webbed neck heart disease, genetic damage of

Short stature wormian bones dextrocardia, genetic damage of

Short syndrome, genetic damage of

Short tarsus absence of lower eyelashes, genetic damage of

Shoulder and thorax deformity congenital heart disease, genetic damage of

Shoulder girdle defect mental retardation familial, genetic damage of

Shprintzen Golberg craniosynostosis, genetic damage of

Shprintzen syndrome, genetic damage of

Shulman syndrome, genetic damage of

Shwachman syndrome, genetic damage of

Shwachman-Diamond syndrome, genetic damage of

Shy-Drager syndrome, genetic damage of

Sialadenitis, genetic damage of

Sialidosis, genetic damage of

Sialidosis type 1 and 3, genetic damage of

Sialuria french type, genetic damage of

Sickle cell anemia, genetic damage of

Sickle cell crisis, genetic damage of

Sickle cell trait, genetic damage of

Sideroblastic anemia, genetic damage of

Siderodromophobia, genetic damage of (possible)

Sidransky Feinstein Goodman syndrome, genetic damage of

Siegler Brewer Carey syndrome, genetic damage of

Silengo Lerone Pelizzo syndrome, genetic damage of

Sillence syndrome, genetic damage of

Silver-Russell dwarfism, genetic damage of

Silvery hair syndrome, genetic damage of

Simosa Penchaszadeh Bustos syndrome, genetic damage of

Simpson-Golabi-Behmel syndrome, genetic damage of

Singh Chhaparwal Dhanda syndrome, genetic damage of

Single upper central incisor, genetic damage of

Single ventricular heart, genetic damage of

Singleton Merten syndrome, genetic damage of

Sinistrophobia, genetic damage of (possible)

Sino-auricular heart block, genetic damage of

Sinus cancer, genetic damage of

Sinus histiocytosis, genetic damage of

Sinus node disease and myopia, genetic damage of

Sipple syndrome, genetic damage of

Sirenomelia, genetic damage of

Sirenomelia sequence, genetic damage of

Sitophobia, genetic damage of (possible)

Sitosterolemia, genetic damage of

Situs inversus viscerum-cardiopathy, genetic damage of

Situs inversus, X linked, genetic damage of

Sjogren Larsson like syndrome, genetic damage of

Sjogren Larsson syndrome, genetic damage of

Sjogren's syndrome, genetic damage of

Skeletal dysplasia brachydactyly, genetic damage of

Skeletal dysplasia epilepsy short stature, genetic damage of

Skeletal dysplasia orofacial anomalies, genetic damage of

Skeletal dysplasia San Diego type, genetic damage of

Skeletal dysplasias, genetic damage of

Skeleto cardiac syndrome with thrombocytopenia, genetic damage of

Sketetal dysplasia coarse facies mental retardation, genetic damage of

Slavotinek Hurst syndrome, genetic damage of

Sleep apnea, genetic damage of

Sly syndrome, genetic damage of

Small cell lung cancer, genetic damage of

Small non-cleaved cell lymphoma, genetic damage of

Small patella syndrome, genetic damage of

Smet Fabry Fryns syndrome, genetic damage of

Smith Fineman Myers syndrome, genetic damage of

Smith Martin Dodd syndrome, genetic damage of

Smith-Magenis syndrome, genetic damage of

Sneddon syndrome, genetic damage of

Sociophobia, genetic damage of (possible)

Soft tissue sarcomas, genetic damage of

Sohval Soffer syndrome, genetic damage of

Somatostatinoma, genetic damage of

Sommer Hines syndrome, genetic damage of

Sommer Rathbun Battles syndrome, genetic damage of

Sommer Young Wee Frye syndrome, genetic damage of

Somniphobia, genetic damage of (possible)

Sondheimer syndrome, genetic damage of

Sonoda syndrome, genetic damage of

Sophophobia, genetic damage of (possible)

Sosby syndrome, genetic damage of

Sotos syndrome, genetic damage of

Sparse hair ptosis mental retardation, genetic damage of

Spasmodic dysphonia, genetic damage of

Spasmodic torticollis, genetic damage of

Spastic angina with healthy coronary artery, genetic damage of

Spastic ataxia Charlevoix-Saguenay type, genetic damage of

Spastic diplegia infantile type, genetic damage of

Spastic dysphonia, genetic damage of

Spastic paraparesis, genetic damage of

Spastic paraparesis deafness, genetic damage of

Spastic paraparesis, infantile, genetic damage of

Spastic paraplegia epilepsy mental retardation, genetic damage of

Spastic paraplegia facial cutaneous lesions, genetic damage of

Spastic paraplegia familial autosomal recessive form, genetic damage of

Spastic paraplegia glaucoma precocious puberty, genetic damage of

Spastic paraplegia mental retardation corpus callosum, genetic damage of

Spastic paraplegia nephritis deafness, genetic damage of

Spastic paraplegia neuropathy poikiloderma, genetic damage of

Spastic paraplegia type 1, X linked, genetic damage of

Spastic paraplegia type 2, X linked, genetic damage of

Spastic paraplegia type 3, dominant, genetic damage of

Spastic paraplegia type 4, dominant, genetic damage of

Spastic paraplegia type 5A, recessive, genetic damage of

Spastic paraplegia type 5B, recessive, genetic damage of

Spastic paraplegia type 6, dominant, genetic damage of

Spastic paraplegia, familial, genetic damage of

Spastic paresis glaucoma mental retardation, genetic damage of

Spastic quadriplegia retinitis pigmentosa mental retardation, genetic damage of

Spasticity mental retardation, genetic damage of

Spasticity multiple exostoses, genetic damage of

Spatic paraparesis vitiligo premature graying, genetic damage of

Spellacy Gibbs Watts syndrome, genetic damage of

Spherocytosis, genetic damage of

Spherophakia brachymorphia syndrome, genetic damage of

Sphingolipidosis, genetic damage of

Sphingomyelinase deficiency, genetic damage of

Spielmeyer-Vogt disease, genetic damage of

Spina bifida, genetic damage of

Spina bifida hypospadias, genetic damage of

Spinal and bulbar muscular atrophy, genetic damage of

Spinal atrophy ophthalmoplegia pyramidal syndrome, genetic damage of

Spinal bulbar motor neuropathy, genetic damage of

Spinal bulbar muscular atrophy, genetic damage of

Spinal cord disorder, genetic damage of

Spinal cord neoplasm, genetic damage of

Spinal dysostosis type Anhalt, genetic damage of

Spinal muscular atrophy, genetic damage of

Spinal muscular atrophy type 1, genetic damage of

Spinal muscular atrophy type 2, genetic damage of

Spinal muscular atrophy type 3, genetic damage of

Spinal muscular atrophy type I with congenital bone fractures, genetic damage of

Spinal stenosis, genetic damage of

Spine rigid cardiomyopathy, genetic damage of

Spinocerebellar ataxia 1, genetic damage of

Spinocerebellar ataxia 2, genetic damage of

Spinocerebellar ataxia 4, genetic damage of

Spinocerebellar ataxia 5, genetic damage of

Spinocerebellar ataxia 6, genetic damage of

Spinocerebellar ataxia 7, genetic damage of

Spinocerebellar ataxia 8, genetic damage of

Spinocerebellar ataxia amyotrophy deafness, genetic damage of

Spinocerebellar ataxia dysmorphism, genetic damage of

Spinocerebellar atrophy type 3, genetic damage of

Spinocerebellar degeneration corneal dystrophy, genetic damage of

Spinocerebellar degenerescence book type, genetic damage of

Spleen neoplasm, genetic damage of

Splenic agenesis syndrome, genetic damage of

Splenogonadal fusion limb defects micrognatia, genetic damage of

Splenomegaly, genetic damage of

Split hand deformity mandibulofacial dysostosis, genetic damage of

Split hand split foot malformation autosomal reces, genetic damage of

Split hand split foot mandibular hypoplasia, genetic damage of

Split hand split foot nystagmus, genetic damage of

Split hand split foot X linked, genetic damage of

Split hand urinary anomalies spina bifida, genetic damage of

Split-hand deformity, genetic damage of

Sponastrime dysplasia, genetic damage of

Spondylarthropathy, genetic damage of

Spondylitis, genetic damage of

Spondylo camptodactyly syndrome, genetic damage of

Spondylo costal dysostosis dandy walker, genetic damage of

Spondylocarpotarsal synostosis, genetic damage of

Spondylocostal dysplasia dominant, genetic damage of

Spondylodysplasia brachyolmia, genetic damage of

Spondyloenchondrodysplasia, genetic damage of

Spondyloepimetaphyseal dysplasia, genetic damage of

Spondyloepimetaphyseal dysplasia congenita, Iraqi type, genetic damage of

Spondyloepimetaphyseal dysplasia congenita, Strudwick type, genetic damage of

Spondyloepimetaphyseal dysplasia joint laxity, genetic damage of

Spondyloepiphyseal dysplasia, genetic damage of

Spondyloepiphyseal dysplasia nephrotic syndrome, genetic damage of

Spondyloepiphyseal dysplasia tarda, genetic damage of

Spondyloepiphyseal dysplasia tarda progressive art, genetic damage of

Spondyloepiphyseal dysplasia, congenital type, genetic damage of

Spondylohypoplasia arthrogryposis popliteal pterygium, genetic damage of

Spondylometaphyseal dysplasia, genetic damage of

Spondylometaphyseal dysplasia Kozlowski type, genetic damage of

Spondylometaphyseal dysplasia, 'corner fracture' type, genetic damage of

Spondylometaphyseal dysplasia, Algerian type, genetic damage of

Spondylometaphyseal dysplasia, Schmidt type, genetic damage of

Spondylometaphyseal dysplasia, Sedaghatian type, genetic damage of

Spondyloperipheral dysplasia short ulna, genetic damage of

Spongy degeneration of central nervous system, genetic damage of

Spontaneous periodic hypothermia, genetic damage of (possible)

Spranger Schinzel Yers syndrome, genetic damage of

Sprengel deformity, genetic damage of

Squamous cell carcinoma, genetic damage of

SSADH (succinic semialdehyde dehydrogenase deficiency), genetic damage of

Stalker Chitayat syndrome, genetic damage of

Stampe Sorensen syndrome, genetic damage of

Stargardt's disease, genetic damage of

Steatocystoma multiplex, genetic damage of

Steatocystoma multiplex natal teeth, genetic damage of

Steele Richardson Olszewski syndrome atypical, genetic damage of

Stein-Leventhal syndrome, genetic damage of

Steinbrocker syndrome, genetic damage of

Steinert disease, genetic damage of

Steinert myotonic dystrophy, genetic damage of

Steinfeld syndrome, genetic damage of

Stenophobia, genetic damage of (possible)

Stern Lubinsky Durrie syndrome, genetic damage of

Sternal cleft, genetic damage of

Sternal cyst vascular anomalies, genetic damage of

Sternal malformation vascular dysplasia associatio, genetic damage of

Steroid dehydrogenase deficiency dental anomalies, genetic damage of

Steroid sulfatase deficiency, genetic damage of

Stevens-Johnson syndrome, genetic damage of

Stickler syndrome, genetic damage of

Stickler syndrome, type 1, genetic damage of

Stickler syndrome, type 2, genetic damage of

Stickler syndrome, type 3, genetic damage of

Stiff man syndrome, genetic damage of

Stiff skin syndrome, genetic damage of

Subaortic stenosis short stature syndrome, genetic damage of

Subcortical laminar heterotopia, genetic damage of

Subependymal nodular heterotopia, genetic damage of

Subpulmonary stenosis, genetic damage of

Subvalvular aortic stenosis, genetic damage of

Succinate coenzyme Q reductase deficiency of, genetic damage of

Succinic acidemia, genetic damage of

Succinic acidemia lactic acidosis congenital, genetic damage of

Succinic semialdehyde dehydrogenase deficiency, genetic damage of

Succinyl-CoA acetoacetate transferase deficiency, genetic damage of

Sucrase-isomaltase deficiency, genetic damage of

Sudden infant death syndrome, genetic damage of

Sugarman syndrome, genetic damage of

Sulfatidosis juvenile, Austin type, genetic damage of

Sulfite and xanthine oxydase deficiency, genetic damage of

Sulfite oxidase deficiency, genetic damage of

Summitt syndrome, genetic damage of

Super mesenteric artery syndrome, genetic damage of

Supranuclear palsy, progressive, genetic damage of

Suriphobia, genetic damage of (possible)

Susac syndrome, genetic damage of

Sutherland Haan syndrome, genetic damage of

Sutton's disease II, genetic damage of

Sweet syndrome, genetic damage of

Swyer syndrome, genetic damage of

Sybert Smith syndrome, genetic damage of

Symmetrical thalamic calcifications, genetic damage of

Symphalangism brachydactyly, genetic damage of

Symphalangism brachydactyly craniosynostosis, genetic damage of

Symphalangism Cushing type, genetic damage of

Symphalangism distal, genetic damage of

Symphalangism familial proximal, genetic damage of

Symphalangism short stature accessory testis, genetic damage of

Symphalangism with multiple anomalies of hands and feet, genetic damage of

Syncamptodactyly scoliosis, genetic damage of

Syncopal paroxysmal tachycardia, genetic damage of

Syncopal tachyarythmia, genetic damage of

Syndactyly, genetic damage of

Syndactyly between 4 and 5, genetic damage of

Syndactyly cataract mental retardation, genetic damage of

Syndactyly Cenani Lenz type, genetic damage of

Syndactyly ectodermal dysplasia cleft lip palate hand foot, genetic damage of

Syndactyly type 1 microcephaly mental retardation, genetic damage of

Syndactyly type 2, genetic damage of

Syndactyly type 3, genetic damage of

Syndactyly type 5, genetic damage of

Syndactyly-polydactyly-ear lobe syndrome, genetic damage of

Syndrome X, genetic damage of

Syngnathia cleft palate, genetic damage of

Syngnathia multiple anomalies, genetic damage of

Synostosis of talus and calcaneus short stature, genetic damage of

Synovial cancer, genetic damage of

Synovial osteochondromatosis, genetic damage of

Synovial sarcoma, genetic damage of

Synovitis, genetic damage of

Synovitis acne pustulosis hyperostosis osteitis, genetic damage of

Synovitis granulomatous uveitis cranial neuropathi, genetic damage of

Synpolydactyly, genetic damage of

Synspondylism, genetic damage of

Syringobulbia, genetic damage of

Syringocystadenoma papilliferum, genetic damage of

Syringomas natal teeth oligodontia, genetic damage of

Syringomelia hyperkeratosis, genetic damage of

Syringomyelia, genetic damage of

Systemic arterio-veinous fistula, genetic damage of

Systemic carnitine deficiency, genetic damage of

Systemic lupus erythematosus, genetic damage of

Systemic mastocytosis, genetic damage of

Systemic sclerosis, genetic damage of

T

T cell immunodeficiency primary, genetic damage of

T-cell lymphoma, genetic damage of

T-Lymphocytopenia, genetic damage of

Tabatznik syndrome, genetic damage of

Tachycardia, genetic damage of

Taeniophobia, genetic damage of (possible)

Takayasu arteritis, genetic damage of

Talipes equinovarus, genetic damage of

Tamari Goodman syndrome, genetic damage of

Tang Hsi Ryu syndrome, genetic damage of

Tangier disease, genetic damage of

Tapinophobia, genetic damage of (possible)

Tar syndrome, genetic damage of

Tardive dyskinesia, genetic damage of

Tarsal tunnel syndrome, genetic damage of

Taste disorder, genetic damage of

Tau syndrome, genetic damage of

Taurodontia absent teeth sparse hair, genetic damage of

Taurodontism, genetic damage of

Tay syndrome ichthyosis, genetic damage of

Tay-Sachs disease, genetic damage of

Taybi Linder syndrome, genetic damage of

Taybi syndrome, genetic damage of

Teebi Kaurah syndrome, genetic damage of

Teebi Naguib Alawadi syndrome, genetic damage of

Teebi Shaltout syndrome, genetic damage of

Teebi syndrome, genetic damage of

Teeth noneruption of with maxillary hypoplasia and genu valgum, genetic damage of

Tel Hashomer camptodactyly syndrome, genetic damage of

Telangiectasia, genetic damage of

Telangiectasia ataxia variant V1, genetic damage of

Telangiectasia, hereditary hemorrhagic, genetic damage of

Telecanthus hypertelorism pes cavus, genetic damage of

Telecanthus with associated abnormalities, genetic damage of

Telencephalic leukoencephalopathy, genetic damage of

Telfer Sugar Jaeger syndrome, genetic damage of

Temporal epilepsy, familial, genetic damage of

Temporomandibular ankylosis, genetic damage of

Temporomandibular joint dysfunction, genetic damage of

Temtamy Shalash syndrome, genetic damage of

TEN, genetic damage of (possible)

Ter Haar Hamel Hendricks syndrome, genetic damage of

Ter Haar syndrome, genetic damage of

Teratocarcinosarcoma, genetic damage of

Teratoma, genetic damage of

Teratophobia, genetic damage of (possible)

Testes neoplasm, genetic damage of

Testicular feminization syndrome, genetic damage of

Testicular regression syndrome, genetic damage of

Testotoxicosis, genetic damage of

Tetanophobia, genetic damage of (possible)

Tethered spinal cord disease, genetic damage of

Tetraamelia ectodermal dysplasia, genetic damage of

Tetraamelia multiple malformations, genetic damage of

Tetraamelia pulmonary hypoplasia, genetic damage of

Tetraamelia-syrinx, genetic damage of

Tetrahydrobiopterin deficiencies, genetic damage of

Tetraploidy, genetic damage of

Tetrasomy 12p, genetic damage of

Tetrasomy 15q, genetic damage of

Tetrasomy 18p, genetic damage of

Tetrasomy 21q, genetic damage of

Tetrasomy 9p, genetic damage of

Tetrasomy X, genetic damage of

Thaasophobia, genetic damage of (possible)

Thakker Donnai syndrome, genetic damage of

Thalamic degeneration symmetrical infantile, genetic damage of

Thalamic degenerescence infantile, genetic damage of

Thalamic syndrome, genetic damage of

Thalassemia, genetic damage of

Thalassemia major, genetic damage of

Thalassemia minor, genetic damage of

Thalassophobia, genetic damage of (possible)

Thanatophobia, genetic damage of (possible)

Thanatophoric dwarfism, genetic damage of

Thanatophoric dysplasia cloverleaf skull, genetic damage of

Thanatophoric dysplasia Glasgow variant, genetic damage of

Thanos Stewart Zonana syndrome, genetic damage of

Theodor Hertz Goodman syndrome, genetic damage of

Thiele syndrome, genetic damage of

Thiemann epiphyseal disease, genetic damage of

Thies Reis syndrome, genetic damage of

Thin ribs tubular bones dysmorphism, genetic damage of

Thiolase deficiency, genetic damage of

Thiopurine S methyltranferase deficiency, genetic damage of

Thomas Jewett Raines syndrome, genetic damage of

Thomas syndrome, genetic damage of

Thombocytopenia X linked, genetic damage of

Thompson Baraitser syndrome, genetic damage of

Thong Douglas Ferrante syndrome, genetic damage of

Thoracic celosomia, genetic damage of

Thoracic dysplasia hydrocephalus syndrome, genetic damage of

Thoracic outlet syndrome, genetic damage of

Thoraco abdominal enteric duplication, genetic damage of

Thoraco limb dysplasia Rivera type, genetic damage of

Thoracolaryngopelvic dysplasia, genetic damage of

Thoracopelvic dysostosis, genetic damage of

Thost-Unna palmoplantar keratoderma, genetic damage of

Thrombasthenia, genetic damage of

Thrombasthenia of Glanzmann and Naegeli, genetic damage of

Thrombocytopathy, genetic damage of

Thrombocytopathy asplenia miosis, genetic damage of

Thrombocytopenia, genetic damage of

Thrombocytopenia cerebellar hypoplasia short stature, genetic damage of

Thrombocytopenia chromosome breakage, genetic damage of

Thrombocytopenia multiple congenital anomaly, genetic damage of

Thrombocytopenia purpura, genetic damage of

Thrombocytopenia Robin sequence, genetic damage of

Thrombocytopenic purpura, autoimmune, genetic damage of

Thrombocytosis, genetic damage of

Thrombomodulin anomalies, familial, genetic damage of

Thrombotic microangiopathy, familial, genetic damage of

Thrombotic thrombocytopenic purpura, genetic damage of

Thumb absence hypoplastic halluces, genetic damage of

Thumb absent short stature immune deficiency, genetic damage of

Thumb deformity alopecia pigmentation anomaly, genetic damage of

Thumb stiff brachydactyly mental retardation, genetic damage of

Thymic epithelial tumor, genetic damage of

Thymic renal anal lung dysplasia, genetic damage of

Thymoma, genetic damage of

Thymus neoplasm, genetic damage of

Thyrocerebrorenal syndrome, genetic damage of

Thyroglossal tract cyst, genetic damage of

Thyroid cancer, genetic damage of

Thyroid carcinoma papillary, genetic damage of

Thyroid carcinoma, follicular, genetic damage of

Thyroid hormone unresponsiveness, genetic damage of

Thyroid renal digital anomalies, genetic damage of

Tibia absent polydactyly, genetic damage of

Tibia absent polydactyly arachnoid cyst, genetic damage of

Tibiae bowed radial anomalies osteopennia fracture, genetic damage of

Tibial aplasia ectrodactyly, genetic damage of

Tibial aplasia ectrodactyly hydrocephalus, genetic damage of

Tibial hemimelia cleft lip palate, genetic damage of

Tibial muscular dystrophy tardive, genetic damage of

Tietze syndrome, genetic damage of

Tinnitus, genetic damage of

TNF-1 receptor associated progressive syndrome, genetic damage of

Todd's paralysis, genetic damage of

Tollner Horst Manzke syndrome, genetic damage of

Tolosa-Hunt syndrome, genetic damage of

Tomaculous neuropathy, genetic damage of

Tome Brune Fardeau syndrome, genetic damage of

Tongue neoplasm, genetic damage of

Toni Debre Fanconi maladie, genetic damage of

Toni-Fanconi syndrome, genetic damage of

Topophobia, genetic damage of (possible)

Toriello Carey syndrome, genetic damage of

Toriello Lacassie Droste syndrome, genetic damage of

Toriello syndrome, genetic damage of

Toriello-Higgins-Miller syndrome, genetic damage of

Torres Ayber syndrome, genetic damage of

Torsion dystonia, genetic damage of

Torticollis keloids cryptorchidism renal dysplasia, genetic damage of

Tosti Misciali Barbareschi syndrome, genetic damage of

Total hypotrichosis, Mari type, genetic damage of

Tourette syndrome, genetic damage of

Townes-Brocks syndrome, genetic damage of

Toxic encephalopathy, genetic damage of

Toxopachyoteose diaphysaire tibio peroniere, genetic damage of

Tracheal agenesis, genetic damage of

Tracheobronchomalacia, genetic damage of

Tracheobronchomegaly, genetic damage of

Tracheobronchopathia osteoplastica, genetic damage of

Tracheoesophageal fistula, genetic damage of

Tracheoesophageal fistula symphalangism, genetic damage of

Tracheophageal fistula hypospadias, genetic damage of

Tranebjaerg Svejgaard syndrome, genetic damage of

Transcobalamin II deficiency, genetic damage of

Transient erythroblastopenia of childhood, genetic damage of

Transient global amnesia, genetic damage of

Transient neonatal arthrogryposis, genetic damage of

Transitional cell carcinoma, genetic damage of

Transposition of great vessels, genetic damage of

Transverse limb deficiency hemangioma, genetic damage of

Transverse myelitits, genetic damage of

TRAPS (TNF-receptor-associated periodic syndrome), genetic damage of

Traumatophobia, genetic damage of (possible)

Treacher Collins-Franceschetti syndrome, genetic damage of

Treft Sanborn Carey syndrome, genetic damage of

Tremophobia, genetic damage of (possible)

Tremor hereditary essential, genetic damage of

Tremor nystagmus duodenal ulcer, genetic damage of

Trevor disease, genetic damage of

Triatrial heart, genetic damage of

Tricho dento osseous syndrome type 1, genetic damage of

Tricho odonto onycho dermal syndrome, genetic damage of

Tricho odonto onychodysplasia syndactyly dominant type, genetic damage of

Tricho onychic dysplasia, genetic damage of

Tricho onycho hypohidrotic dysplasia, genetic damage of

Tricho retino dento digital syndrome, genetic damage of

Tricho-dento-osseous syndrome, genetic damage of

Tricho-hepato-enteric syndrome, genetic damage of

Trichodental syndrome, genetic damage of

Trichodermal syndrome mental retardation, genetic damage of

Trichodermodysplasia dental alterations, genetic damage of

Trichodysplasia xeroderma, genetic damage of

Trichoepithelioma multiple familial, genetic damage of

Trichofolliculloma, genetic damage of

Trichomalacia, genetic damage of

Trichomegaly cataract hereditary spherocytosis, genetic damage of

Trichomegaly retina pigmentary degeneration dwarfism, genetic damage of

Trichoodontoonychial dysplasia, genetic damage of

Trichopathophobia, genetic damage of (possible)

Trichorhinophalangeal syndrome type I, genetic damage of

Trichorhinophalangeal syndrome type II, genetic damage of

Trichorhinophalangeal syndrome type III, genetic damage of

Trichostasis spinulosa, genetic damage of

Trichothiodystrophy (generic term), genetic damage of

Trichothiodystrophy sun sensitivity, genetic damage of

Trichothiodystrophy with congenital ichtyosis, genetic damage of

Tricuspid atresia, genetic damage of

Tricuspid dysplasia, genetic damage of

Trigonocephaly bifid nose acral anomalies, genetic damage of

Trigonocephaly broad thumbs, genetic damage of

Trigonocephaly ptosis coloboma, genetic damage of

Trigonocephaly ptosis mental retardation, genetic damage of

Trigonomacrocephaly tibial defect polydactyly, genetic damage of

Trihydroxycholestanoylcoa oxidase isolated deficiency, genetic damage of

Triopia, genetic damage of

Triose phosphate-isomerase deficiency, genetic damage of

Triphalangeal thumb non-opposable, genetic damage of

Triphalangeal thumb polysyndactyly syndrome, genetic damage of

Triphalangeal thumbs brachyectrodactyly, genetic damage of

Triple A syndrome, genetic damage of

Triplo X syndrome, genetic damage of

Triploid syndrome, genetic damage of

Triploidy, genetic damage of

Trismus pseudocamptodactyly syndrome, genetic damage of

Trisomy, genetic damage of

Trisomy 1 mosaicism, genetic damage of

Trisomy 10p, genetic damage of

Trisomy 10pter p13, genetic damage of

Trisomy 10q, genetic damage of

Trisomy 10q partial, genetic damage of

Trisomy 11q, genetic damage of

Trisomy 11q23, genetic damage of

Trisomy 12 mosaicism, genetic damage of

Trisomy 12p, genetic damage of

Trisomy 13, genetic damage of

Trisomy 13 syndrome, genetic damage of

Trisomy 13p, genetic damage of

Trisomy 13q, genetic damage of

Trisomy 14 mosaicism, genetic damage of

Trisomy 14qprox, genetic damage of

Trisomy 14qter, genetic damage of

Trisomy 15 mosaicism, genetic damage of

Trisomy 15q, genetic damage of

Trisomy 16 (mosaic trisomy 16), genetic damage of

Trisomy 16p, genetic damage of

Trisomy 16q, genetic damage of

Trisomy 17 mosaicism, genetic damage of

Trisomy 17p, genetic damage of

Trisomy 17p11 2, genetic damage of

Trisomy 17q22, genetic damage of

Trisomy 18, genetic damage of

Trisomy 18 mosaicism, genetic damage of

Trisomy 18 syndrome, genetic damage of

Trisomy 18p, genetic damage of

Trisomy 18q, genetic damage of

Trisomy 19q, genetic damage of

Trisomy 1p21 p32, genetic damage of

Trisomy 1q32 qter, genetic damage of

Trisomy 1q42 11 q42 12, genetic damage of

Trisomy 1q42 qter, genetic damage of

Trisomy 2 mosaicism, genetic damage of

Trisomy 20 mosaicism, genetic damage of

Trisomy 20p, genetic damage of

Trisomy 21, genetic damage of

Trisomy 22, genetic damage of

Trisomy 22q11 q13, genetic damage of

Trisomy 2p, genetic damage of

Trisomy 2p13 p21, genetic damage of

Trisomy 2pter p24, genetic damage of

Trisomy 2q, genetic damage of

Trisomy 2q37, genetic damage of

Trisomy 3 mosaicism, genetic damage of

Trisomy Partial 8, genetic damage of

Trisomy Xp3, genetic damage of

Trisomy Xpter Xq13, genetic damage of

Trisomy Xq, genetic damage of

Trisomy Xq25, genetic damage of

Trochlear dysplasia, genetic damage of

Trophoblastic neoplasms (gestational trophoblastic disease), genetic damage of

Trophoblastic tumor, genetic damage of

Tropical spastic paraparesis, genetic damage of

Tropophobia, genetic damage of (possible)

Troyer syndrome, genetic damage of

True hermaphroditism, genetic damage of

Trueb Burg Bottani syndrome, genetic damage of

Trypanophobia, genetic damage of (possible)

Tsao Ellingson syndrome, genetic damage of

Tsukahara Azuno Kajii syndrome, genetic damage of

Tsukahara Kajii syndrome, genetic damage of

Tsukuhara syndrome, genetic damage of

Tuberous sclerosis, genetic damage of

Tuberous sclerosis, type 1, genetic damage of

Tuberous sclerosis, type 2, genetic damage of

Tucker syndrome, genetic damage of

Tuffli Laxova syndrome, genetic damage of

Tufted angioma, genetic damage of (possible)

Tunglang Savage Bellman syndrome, genetic damage of

Turcot syndrome, genetic damage of

Turner Kieser syndrome, genetic damage of

Turner Morgani Albright, genetic damage of

Turner phenotype with normal karyotype, genetic damage of

Turner's syndrome, genetic damage of

Turner-like syndrome, genetic damage of

Tutuncuoglu syndrome, genetic damage of

Tyrosine transaminase deficiency, genetic damage of

Tyrosine-oxidase temporary deficiency, genetic damage of

Tyrosinemia, genetic damage of

Tyrosinemia type 1, genetic damage of

Tyrosinemia type 2, genetic damage of

U

UDD tibial myopathy, genetic damage of

UDP-galactose-4-epimerase deficiency, genetic damage of

Uhl anomaly, genetic damage of

Ulbright Hodes syndrome, genetic damage of

Ulcerative colitis, genetic damage of (possible)

Ulerythema ophryogenesis, genetic damage of

Ulna and fibula absence with severe limb deficit, genetic damage of

Ulna hypoplasia, genetic damage of

Ulna hypoplasia mental retardation, genetic damage of

Ulna metaphyseal dysplasia syndrome, genetic damage of

Ulnar hypoplasia lobster claw deformity of feet, genetic damage of

Ulnar mammary syndrome, genetic damage of

Ulnar mammary syndrome of Pallister, genetic damage of

Umbilical cord ulceration intestinal atresia, genetic damage of

Uncombable hair syndrome, genetic damage of

Uniparental disomy, genetic damage of

Uniparental disomy of 10, genetic damage of

Uniparental disomy of 11, genetic damage of

Uniparental disomy of 13, genetic damage of

Uniparental disomy of 14, genetic damage of

Uniparental disomy of 15, genetic damage of

Uniparental disomy of 16, genetic damage of

Uniparental disomy of 2, genetic damage of

Uniparental disomy of 21, genetic damage of

Uniparental disomy of 22, genetic damage of

Uniparental disomy of 5, genetic damage of

Uniparental disomy of 6, genetic damage of

Uniparental disomy of 7, genetic damage of

Uniparental disomy of 8, genetic damage of

Uniparental disomy of 9, genetic damage of

Upington disease, genetic damage of

Upper limb defect eye and ear abnormalities, genetic damage of

Upton Young syndrome, genetic damage of

Urachal cancer, genetic damage of

Urachal cyst, genetic damage of

Urban Rogers Meyer syndrome, genetic damage of

Urban Schosser Spohn syndrome, genetic damage of

Urea cycle enzymopathies, genetic damage of

Uremia, genetic damage of

Urethral obstruction sequence, genetic damage of

Uridine monophosphate synthetase deficiency, genetic damage of

Urinary calculi, genetic damage of (possible)

Urinary tract neoplasm, genetic damage of

Urioste Martinez Frias syndrome, genetic damage of

Urogenital adysplasia, genetic damage of

Urophathy distal obstructive polydactyly, genetic damage of

Urticaria, genetic damage of

Urticaria pigmentosa, genetic damage of

Urticaria-deafness-amyloidosis, genetic damage of

Usher syndrome, genetic damage of

Usher syndrome, type 1A, genetic damage of

Usher syndrome, type 1B, genetic damage of

Usher syndrome, type 1C, genetic damage of

Usher syndrome, type 1D, genetic damage of

Usher syndrome, type 1E, genetic damage of

Usher syndrome, type 2A, genetic damage of

Usher syndrome, type 2B, genetic damage of

Usher syndrome, type 3, genetic damage of

Uveal diseases, genetic damage of

Uveitis, genetic damage of

Uveitis, anterior, genetic damage of

Uveitis, posterior, genetic damage of

V

VACTERL association, genetic damage of

Vacterl hydrocephaly, genetic damage of

Vacuolar myopathy, genetic damage of

Vagina, absence of, genetic damage of

Vagneur Triolle Ripert syndrome, genetic damage of

Valinemia, genetic damage of

Valvular dysplasia of the child, genetic damage of

Van Allen Myhre syndrome, genetic damage of

Van Bogaert disease, genetic damage of

Van Den Berghe Dequeker syndrome, genetic damage of

Van Den Bosch syndrome, genetic damage of

Van Den Ende Brunner syndrome, genetic damage of

Van der Woude syndrome, genetic damage of

Van Goethem syndrome, genetic damage of

Van Maldergem Wetzburger Verloes syndrome, genetic damage of

Van Regemorter Pierquin Vamos syndrome, genetic damage of

Varadi Papp syndrome, genetic damage of

Variegate porphyria, genetic damage of

Vas deferens, congenital bilateral aplasia of, genetic damage of

Vascular disruption sequence, genetic damage of

Vascular malformations of the brain, genetic damage of

Vascular malposition, genetic damage of

Vascular purpura, genetic damage of

Vasculitis hypersensitivity, genetic damage of

Vasculitis, cutaneous necrotizing, genetic damage of

Vasopressin-resistant diabetes insipidus, genetic damage of

Vasquez Hurst Sotos syndrome, genetic damage of

VATER association, genetic damage of

Vein of Galen aneurysm, genetic damage of

Velocardiofacial syndrome, genetic damage of

Velofacioskeletal syndrome, genetic damage of

Velopharyngeal incompetence, genetic damage of

Venencie Powell Winkelmann syndrome, genetic damage of

Ventricular extrasystoles perodactyly Robin sequence, genetic damage of

Ventricular familial preexcitation syndrome, genetic damage of

Ventricular fibrillation, idiopathic, genetic damage of

Ventricular septal defects, genetic damage of

Ventriculo-arterial discordance, isolated, genetic damage of

Ventruto Digirolamo Festa syndrome, genetic damage of

Venustraphobia, genetic damage of (possible)

Verbophobia, genetic damage of (possible)

Verloes Bourguignon syndrome, genetic damage of

Verloes David syndrome, genetic damage of

Verloes Gillerot Fryns syndrome, genetic damage of

Verloes Van Maldergem Marneffe syndrome, genetic damage of

Verloove Vanhorick Brubakk syndrome, genetic damage of

Verminiphobia, genetic damage of (possible)

Vernal keratoconjunctivitis, genetic damage of

Verneuil disease, genetic damage of

Verrucous nevus, genetic damage of

Verrucous nevus acanthokeratolytic, genetic damage of

Vertebral body fusion overgrowth, genetic damage of

Vertebral fusion posterior lumbosacral blepharoptosis, genetic damage of

Vertical talus, genetic damage of

Vestibulocochlear dysfunction progressive familial, genetic damage of

Vestiphobia, genetic damage of (possible)

Viljoen Kallis Voges syndrome, genetic damage of

Viljoen Smart syndrome, genetic damage of

Viljoen Winship syndrome, genetic damage of

Vipoma, genetic damage of

Virginitiphobia, genetic damage of (possible)

Virilism, genetic damage of

Virilizing ovarian tumor, genetic damage of

Virus associated hemophagocytic syndrome, genetic damage of (possible)

Visceral myopathy familial external ophthalmoplegia, genetic damage of

Viscero-atrial heterotaxia, genetic damage of

Vitamin B12 responsive methylmalonic acidemia, cbl A, genetic damage of

Vitamin B12 responsive methylmalonicaciduria, genetic damage of

Vitamin D resistant rickets, genetic damage of

Vitiligo, genetic damage of

Vitiligo mental retardation facial dysmorphism uremia, genetic damage of

Vitiligo psychomotor retardation cleft palate facial dysmorphism, genetic damage of

Vitreoretinal degeneration, genetic damage of

Vitreoretinochoroidopathy dominant, genetic damage of

VKH, genetic damage of

VLCAD deficiency, genetic damage of

Vocal cord dysfunction familial, genetic damage of

Vohwinkel syndrome, genetic damage of

Von Gierke disease, genetic damage of

Von Recklinghausen disease, genetic damage of

Von Voss Cherstvoy syndrome, genetic damage of

Von Willebrand disease, genetic damage of

Von Willebrand disease, dominant form, genetic damage of

Von Willebrand disease, recessive form, genetic damage of

Vulvar vestibulitis syndrome, genetic damage of

Vulvoldynia, genetic damage of (possible)

W

W syndrome, genetic damage of

Waaler Aarskog syndrome, genetic damage of

Waardenburg syndrome, genetic damage of

Waardenburg syndrome type 1, genetic damage of

Waardenburg syndrome type 2, genetic damage of

Waardenburg syndrome type 2A, genetic damage of

Waardenburg syndrome type 2B, genetic damage of

Waardenburg syndrome, type 3, genetic damage of

Waardenburg syndrome, type 4, genetic damage of

Waardenburg type Pierpont, genetic damage of

Waardenburg-Shah syndrome, genetic damage of

Wagner disease, genetic damage of

Wagner-Stickler syndrome, genetic damage of

WAGR syndrome, genetic damage of

Walbaum Titran Durieux Crepin syndrome, genetic damage of

Waldenstrom macroglobulinemia, genetic damage of

Waldmann disease, genetic damage of

Walker Dyson syndrome, genetic damage of

Wallerian degeneration, genetic damage of

Wallis Zieff Goldblatt syndrome, genetic damage of

Wandering spleen, genetic damage of

Warburg Sjo Fledelius syndrome, genetic damage of

Warburg Thomsen syndrome, genetic damage of

Warburton Anyane Yeboa syndrome, genetic damage of

Warkany, genetic damage of

Warm-reacting-antibody hemolytic anemia, genetic damage of

Warman Mulliken Hayward syndrome, genetic damage of

Watermelon stomach, genetic damage of

Watson Alagille syndrome, genetic damage of

Watson syndrome, genetic damage of

Weaver Johnson syndrome, genetic damage of

Weaver like syndrome, genetic damage of

Weaver syndrome, genetic damage of

Weaver Williams syndrome, genetic damage of

Weber Parkes syndrome, genetic damage of

Weber Sturge Dimitri syndrome, genetic damage of

Weber-Christian disease, genetic damage of

Webster Deming syndrome, genetic damage of

Wegener's granulomatosis, genetic damage of

Wegmann Jones Smith syndrome, genetic damage of

Weil syndrome, genetic damage of

Weill-Marchesani syndrome, genetic damage of

Weinstein Kliman Scully syndrome, genetic damage of

Weismann Netter Stuhl syndrome, genetic damage of

Weismann Netter syndrome, genetic damage of

Weissenbacher Zweymuller syndrome, genetic damage of

Welander distal myopathy, Swedish type, genetic damage of

Weleber Hecht Bigley syndrome, genetic damage of

Wellesley Carmen French syndrome, genetic damage of

Wells Jankovic syndrome, genetic damage of

Wells syndrome, genetic damage of

Werdnig-Hoffmann disease, genetic damage of

Werner's syndrome, genetic damage of

Wernicke Korsakoff syndrome, genetic damage of

West syndrome, genetic damage of

Westerhof Beemer Cormane syndrome, genetic damage of

Western equine encephalitis, genetic damage of

Westphall disease, genetic damage of

Whipple disease, genetic damage of

Whitaker syndrome, genetic damage of

White forelock with malformations, genetic damage of

White matter hypoplasia corpus callosum agenesia mental retardation, genetic damage of

Whyte Murphy syndrome, genetic damage of

Wiccaphobia, genetic damage of (possible)

Wieacker syndrome, genetic damage of

Wieacker-Wolff syndrome, genetic damage of

Wiedemann Grosse Dibbern syndrome, genetic damage of

Wiedemann Oldigs Oppermann syndrome, genetic damage of

Wiedemann Opitz syndrome, genetic damage of

Wiedemann Rautenstrauch syndrome, genetic damage of

Wildervanck syndrome, genetic damage of

Wilkes Stevenson syndrome, genetic damage of

Wilkie Taylor Scambler syndrome, genetic damage of

Willebrand disease type 1, genetic damage of

Willebrand disease type 2A, genetic damage of

Willebrand disease type 2B, genetic damage of

Willebrand disease type 2M, genetic damage of

Willebrand disease type 2N, genetic damage of

Willebrand disease type 3, genetic damage of

Willems de Vries syndrome, genetic damage of

Willi-Prader syndrome, genetic damage of

Williams syndrome, genetic damage of

Wilms tumor and pseudohermaphroditism, genetic damage of

Wilms tumor-aniridia, genetic damage of

Wilms tumour radial bilateral aplasia, genetic damage of

Wilms' tumor, genetic damage of

Wilson disease, genetic damage of

Wilson Turner syndrome, genetic damage of

Winchester syndrome, genetic damage of

Winkelman Bethge Pfeiffer syndrome, genetic damage of

Winship Viljoen Leary syndrome, genetic damage of

Winter Harding Hyde syndrome, genetic damage of

Winter Shortland Temple syndrome, genetic damage of

Wisconsin syndrome, genetic damage of

Wiskott Aldrich syndrome, genetic damage of

Witkop syndrome, genetic damage of

Wohlwill-Andrade syndrome, genetic damage of

Wolcott-Rallison syndrome, genetic damage of

Wolf-Hirschorn syndrome, genetic damage of

Wolff Zimmermann syndrome, genetic damage of

Wolff-Parkinson-White syndrome, genetic damage of

Wolfram syndrome, genetic damage of

X

X linked mental retardation type Schutz, genetic damage of

X linked mental retardation type Snyder, genetic damage of

X linked mental retardation type Wittner, genetic damage of

X linked alpha thalassemia mental retardation syndrome (ATR-X), genetic damage of

X linked dominant, genetic damage of

X linked ichthyosis, genetic damage of

X linked juvenile retinoschisis, genetic damage of

X linked lymphoproliferative syndrome, genetic damage of

X linked mental retardation and macro-orchidism, genetic damage of

X linked mental retardation associated with marXq2, genetic damage of

X linked mental retardation-hypotonia, genetic damage of

X linked recessive, genetic damage of

X linked severe combined immunodeficiency, genetic damage of

Xanthic urolithiasis, genetic damage of

Xanthine oxydase deficiency, genetic damage of

Xanthinuria, genetic damage of

Xanthomatosis cerebrotendinous, genetic damage of

Xanthophobia, genetic damage of (possible)

Xenoglossophobia, genetic damage of (possible)

Xerocytosis, hereditary, genetic damage of

Xeroderma pigmentosum, genetic damage of

Xeroderma pigmentosum, type 1, genetic damage of

Xeroderma pigmentosum, type 2, genetic damage of

Xeroderma pigmentosum, type 3, genetic damage of

Xeroderma pigmentosum, type 5, genetic damage of

Xeroderma pigmentosum, type 6, genetic damage of

Yunis Varon syndrome, genetic damage of

Z

Zadik Barak Levin syndrome, genetic damage of

ZAP70 deficiency, genetic damage of

Zazam Sheriff Phillips syndrome, genetic damage of

Zellweger syndrome, genetic damage of

Zimmerman Laband syndrome, genetic damage of

Zlotogora syndrome, genetic damage of

Zollinger-Ellison syndrome, genetic damage of

Zoophobia, genetic damage of (possible)

Zori Stalker Williams syndrome, genetic damage of

Zunich-Kaye syndrome, genetic damage of

Zuska's disease, genetic damage of

8. A Catalogue of Genetic Damage

"All" of us know the DSM—IV—TR (Diagnostic and Statistical Manual of Mental Disorders). But what is DSM—IV—TR, basically? The index does not contain words like "gene damage", "genetic damage", "genetic disease", or "genetic disorder". I do not think the authors of this book do really fully understand what they are writing about. The fact is that DSM—IV—TR is a catalogue of genetic damage to the human instincts (the basic elements of human behavior). Undoubtedly, basically the human mental disorders are gene-damaged human instincts.

9. Genetic Damage in Dogs

[COPY FROM THE INTERNET:]

Canine Consumer Report
A Guide to Hereditary and Congenital Diseases in Dogs*

(*Includes Genetic Predisposition to Diseases)

Published by The Association of Veterinarians for Animal Rights, P.O. Box 208, Davis, CA 95617-0208
First printing: August 1994
Special thanks to: W. Jean Dodds, D.V.M., Susan Hall, D.V.M., Kay Inks, and Phil Maggitti

Directory
Section I

A list of 148 purebred dog breeds, each of which is accompanied by a number or series of numbers that correspond to the genetic diseases identified and described in Section II.'

Section II

An alphabetical listing of genetically transmitted diseases that occur in purebred dogs. Each disease is assigned an identification number, and some diseases are followed by the names of the breeds known to be subject to those diseases.
How to use this book
If you are thinking about acquiring a purebred dog, refer to Section I to find out the number of genetically transmitted diseases associated with the breed or breeds in which you are interested. Refer to Section II to find the names and definitions of those diseases.

Disclaimer

This report was designed to catalogue the existing congenital and hereditary diseases found in purebred dogs and to describe other common conditions that are thought to be genetic in origin because they appear with higher-than-expected frequency in certain breeds. The knowledge of and experience with nonhuman animal diseases is always changing, and new diseases are discovered each year. As a result, this catalogue cannot be fully complete. Every attempt has been made, nevertheless, to provide the latest information from published articles in scientific and popular journals, from review chapters written by veterinary and other professionals, and from breed-club literature provided by designated representatives of individual breed clubs. The conditions listed in the Canine Consumer Report occur with varying degrees of frequency and severity from one breed to the next. Hip dysplasia, umbilical hernias, and allergies, for example, are common conditions found in many breeds, while serious malformations (such as the lysosomal storage diseases) and life-threatening diseases (such as hemophilia and certain cancers) are found with less frequency. The impact of a particular disorder on the health and longevity of the breed or breeds that it affects will depend on the nature and severity of that disorder. Therefore, no attempt has been made to assign priority among the diseases identified in this catalogue. This sort of information should be sought form veterinarians and other medical professionals with relevant experience and from national breed-club representatives, whose names can be obtained from the American Kennel Club, 51 Madison Avenue, New York, NY 10010.

Section I:

Aberdeen terrier: 324
Affenpinscher: 12, 55, 98, 218, 235, 236, 330
Afghan hound: 14, 42, 65, 96, 114, 121, 135, 145, 147, 166, 206a, 211, 221, 221a, 239, 245, 256, 269, 270, 330
Airedale terrier: 7, 9, 48, 65, 88, 103, 122, 140, 148, 165, 166, 168, 206a, 230, 256, 269, 270, 273a, 314, 318, 330
Akita: 9, 27, 43, 65, 71, 103, 114, 115, 135, 137, 140, 146, 152, 156, 166, 172,192, 193a, 199, 221, 221a, 239, 256, 270, 273a, 312, 318, 329, 330
Alaskan malamute: 13, 42, 59, 65, 67, 85, 89, 120, 121, 122, 135, 144, 147, 148, 150, 152, 166, 206a, 221, 221a, 256, 266, 330, 334
American cocker spaniel: 1, 10, 12, 18, 26, 27, 42, 43, 54, 55, 65, 69, 72, 73, 88, 94, 94a, 95, 103, 107, 109, 121, 123, 135, 146, 147, 148, 152, 154, 166, 171,

173, 179, 181, 186, 188, 193a, 197, 221, 221a, 226, 228, 235, 236, 242, 245, 254, 256, 266, 270, 275, 276, 286, 307, 318, 319, 320, 330
American foxhound: 78, 199, 290, 311
American Staffordshire terrier: 42, 54, 55, 88, 103, 166, 204a, 221, 221a, 256
American water spaniel: 42, 150, 270
Antarctic husky: 103, 147
Australian cattle dog (Australian blue heeler): 42, 78, 109, 121, 124b, 147, 149, 152, 166, 171, 177, 186, 193a, 200, 214, 221, 228, 245, 256, 270, 318, 319
Australian kelpie: 58, 199, 203, 256
Australian shepherd: 42, 55, 58, 78, 89, 152, 166, 199, 203, 221, 221a, 245, 256, 269, 270, 287, 318, 328, 329, 330
Australian terrier: 85, 185, 256, 270
Basenji: 56, 59, 66, 124a, 146, 166, 171, 172, 245, 256, 263, 268, 270, 318
Basset hound: 5, 9, 9a, 15, 27, 31, 61a, 70, 94, 103, 105, 109, 114, 120, 121, 131, 135, 136, 146, 147, 157, 159a, 166, 168, 169, 170, 171, 174, 186, 190, 221, 221a, 222, 231, 235, 245, 249, 256, 274, 291, 299, 311, 318, 330, 332
Beagle: 10, 11, 21, 34, 37, 42, 43, 54, 55, 65, 72, 80, 88, 94a, 109, 114, 120, 121, 135, 136, 146, 147, 157, 166, 168, 173, 182, 188a, 192, 193a, 202, 204, 212, 220, 227, 242, 245, 256, 260, 267, 270, 275, 280, 310, 312, 327, 330
Bearded collie: 9a, 42, 65, 152, 159a, 166, 245, 256, 269, 270, 303
Bedlington terrier: 2, 23, 42, 88, 94, 184, 199, 210, 223, 256, 265, 266, 269, 270
Belgian malinois: 109, 152, 166, 256
Belgian sheepdog: 42, 109, 152, 166, 204a, 230, 256, 269, 270
Belgian tervuren: 42, 109, 166, 221, 221a, 230, 256
Bernese mountain dog: 20, 42, 46, 47, 54, 55, 103, 149, 149d, 152, 166, 221, 221a, 255, 256, 269, 318
Bichon frise: 42, 53, 65, 103, 109, 122, 148, 235, 331a
Black and tan coonhound: 94, 103, 122, 148, 152, 221, 221a, 252
Bloodhound: 31, 94, 103, 114, 152, 166, 179, 181, 195, 221, 221a, 245, 324
Blue tick hound: 136, 193a, 221, 221a
Border collie: 65, 109, 152, 186, 214, 221, 221a, 256
Border terrier: 17, 39, 42, 58, 68, 70, 126, 145, 186, 196, 217, 235, 248, 256, 270, 324, 327
Borzoi (Russian wolfhound): 31, 36, 42, 118, 152, 155, 166, 192, 200, 256, 312, 330

199, 206a, 221, 221a, 231, 243, 245, 250, 251, 256, 266, 266a, 267, 270, 292, 304a, 312, 328, 330

English bulldog: 1, 3, 6, 19, 42, 54, 55, 80, 88, 90, 94, 98, 103, 116, 129, 130, 145, 152, 154, 164, 166, 179, 181, 201, 205, 217, 242, 245, 260, 261, 278, 280, 287, 308, 325, 330

English cocker spaniel: 42, 70, 88, 94, 103, 119, 135, 147, 150, 166, 177, 186, 214, 221, 221a, 236, 245, 256, 259, 270, 304, 330

English foxhound: 78, 290

English setter: 31, 42, 61, 68, 78, 91, 94, 103, 121, 124b, 147, 152, 160, 166, 177, 181, 188a, 193a, 214, 221, 221a, 256, 323, 330

English springer spaniel: 9a, 10, 12, 18, 26, 27, 42, 43, 54, 55, 59, 65, 69, 72, 88, 94, 94a, 95, 103, 107, 109, 121, 123, 124, 135, 146, 147, 148, 152, 154, 159a, 160a, 166, 171, 173, 188, 193a, 197, 206a, 220, 221, 221a, 226, 228, 235, 236, 242, 245, 245a, 254, 256, 264a, 266, 270, 275, 276, 286, 307, 318, 319, 320, 330

English toy spaniel (King Charles and Ruby Blenheim spaniels): 42, 55, 65, 85, 103, 143, 235, 270, 318

Field spaniel: 14, 42, 166, 256, 270

Finnish spitz: 85

Flat-coated retriever: 42, 88, 94, 103, 152, 166, 256

French bulldog: 42, 54, 55, 88, 98, 103, 119, 121, 122, 145, 147, 148, 330

German shepherd: 10, 21, 27, 36, 42, 44, 54, 55, 59, 65, 72, 72a, 75, 81, 83, 86, 94a, 95, 102, 103, 105, 109, 112, 114, 121, 122, 129a, 221a, 131, 137, 143a, 147, 148, 152, 166, 168, 180, 186, 191, 192, 193a, 194, 208, 220, 221, 221a, 225, 226, 229, 230, 231, 236, 238, 241, 243, 247, 256, 266, 270, 276, 283, 300, 306, 312, 316, 320, 330

German short-haired pointer: 8, 9a, 42, 65, 103, 114, 124b, 125, 152, 159a, 166, 177, 188a, 191, 193a, 197, 214, 221, 221a, 230, 256, 300, 311, 330

German wire-haired pointer: 42, 103, 152, 221, 221a, 270, 302, 330

Giant schnauzer: 42, 105, 135, 146, 152, 158, 166, 192, 221, 221a, 231, 256, 269, 270, 276, 311a, 312, 327a

Golden retriever: 7, 9, 10, 22, 42, 59, 65, 81, 88, 94, 95, 103, 121, 129, 129a, 130, 140, 143a, 146, 147, 149, 152, 166, 178, 192, 193, 204a, 206, 220, 221, 221a, 256, 273a, 300, 312, 328, 329

Gordon setter: 31, 42, 45, 103, 152, 166, 221, 221a, 256, 179, 181, 270, 307

Great Dane: 6, 7, 31, 36, 38, 42, 48, 50, 51, 61, 75, 78, 80, 83, 88, 94, 103, 114, 131, 135, 144, 152, 153, 155, 158, 166, 176, 192, 198, 199, 201, 204a, 211, 221, 221a, 225, 243, 250, 255, 256, 270, 292, 298, 312, 330

Miniature poodle: 5, 10, 22, 26, 27, 42, 49, 78, 81, 88, 92, 93, 103, 109, 110, 111, 121, 135, 136, 140, 144, 147, 156, 165, 166, 173, 175, 184, 193a, 199, 206, 220, 221, 221a, 223, 226, 230, 235, 236, 256, 269, 275, 294, 327, 330

Miniature schnauzer: 10, 22, 42, 70, 76, 88, 103, 112, 121, 144a, 146, 147, 149, 149a, 157, 166, 185, 221, 221a, 259, 260, 266a, 271, 284, 301, 311a, 330

Neapolitan mastiff: 42, 83, 94, 103, 152, 158, 166, 221, 245, 256

Newfoundland: 25, 38, 42, 75, 83, 94, 95, 103, 114, 129a, 146, 152, 166, 183, 192, 221, 221a, 236, 300, 311a, 312, 320, 327, 330

Norwegian dunkerhound: 78, 199,

Norwegian elkhound: 42, 88, 103, 135, 152, 166, 180, 186, 256, 266, 275, 276, 302

Norwich terrier: 65, 166, 186, 330

Nova Scotia duck tolling retriever: 9a, 42, 159a, 166, 256

Old English sheepdog: 9a, 42, 80, 88, 103, 122, 129, 140, 146, 148, 152, 159a, 161, 166, 172, 192, 221, 221a, 250, 256, 269, 270, 273a, 292, 311a, 312, 328, 330

Otter hound: 119, 152, 166, 221, 221a, 249, 274, 311, 330

Papillon: 12, 42, 65, 103, 166, 235, 330

Pekingese: 42, 88, 103, 116, 146, 162, 166, 171, 173, 179, 181, 184, 186, 199, 230, 246, 256, 277, 278, 304, 311a, 317, 318

Pembroke Welsh corgi: 42, 50, 65, 72, 75, 90, 109, 166, 186, 245, 256, 270, 330

Pharoah hound: 10, 166, 220, 311a

Pointer: 8, 10, 36, 42, 65, 80, 89, 103, 109, 152, 166, 178, 213, 215, 230, 231, 239, 256, 318, 330

Pomeranian: 42, 70, 73, 87, 88, 103, 110, 136, 137, 140, 149, 162, 165, 166, 184, 186, 210, 219, 235, 236, 256, 273a, 313

Portuguese water dog: 9a, 42, 88, 159a, 166, 188a, 193a, 245, 256, 273a, 299a

Pug: 10, 22, 50, 54, 57, 65, 67, 76, 80, 81, 90, 98, 98a, 103, 109, 116, 143, 149, 152, 166, 173, 179, 185, 195, 196, 230, 235, 246, 256, 259, 293, 295, 304a, 308, 309, 317

Puli: 27, 42, 152, 256, 270

Rhodesian ridgeback: 42, 45, 51, 84, 103, 152, 166, 221, 221a, 245, 256

Rottweiler: 9a, 27, 42, 85, 88, 94, 95, 103, 105, 129, 129a, 146, 152, 159a, 161, 166, 172, 192, 231, 245, 256, 269, 270, 300, 311a, 312, 326, 328, 330

Saint Bernard: 31, 42, 60, 72, 83, 88, 94, 94a, 103, 109, 114, 118, 121, 122, 133, 147, 148, 149, 152, 166, 188, 198, 221, 221a, 225, 298, 325, 328, 329, 330

Saluki: 14, 27, 42, 65, 103, 146, 166, 214, 245, 256, 269, 270, 311a, 330
Samoyed: 24, 42, 65, 85, 88, 89, 103, 121, 135, 140, 147, 149, 149b, 152, 166, 221, 221a, 242, 245, 256, 260, 269, 270, 273a, 274, 311a, 328, 330
Schipperke: 42, 85, 88, 103, 166, 185, 207, 239, 245, 256
Scottish deerhound: 31, 42, 105, 131, 158, 166, 221, 221a
Scottish terrier: 5, 10, 22, 42, 68, 75, 78, 81, 122, 129, 148, 166, 186, 193, 197, 256, 272, 294, 324, 330
Sealyham terrier: 22, 42, 81, 135, 166, 186, 245, 256, 269, 270
Shetland sheepdog: 5, 42, 52, 59, 65, 82, 86, 88, 108, 121, 122, 129, 147, 148, 149a, 151, 152, 157, 166, 192, 220, 236, 245, 256, 270, 306, 312, 328, 329, 330
Shih Tzu: 1, 42, 54, 55, 83, 88, 89, 94, 103, 146, 149, 166, 182, 187, 256, 266, 269, 311a, 317, 330
Siberian husky: 41, 42, 65, 86, 103, 104, 121, 135, 147, 152, 166, 186, 221, 221a, 230, 245, 256, 270, 328, 329, 330, 334
Silky terrier: 42, 70, 85, 154, 185, 188a, 193a, 235, 245, 256, 311a, 313
Skye terrier: 27, 88, 101, 111, 163, 183, 186, 192, 206, 312, 316, 330
Smooth fox terrier: 3, 10, 22, 42, 78, 81, 87, 88, 112, 135, 138, 166, 185, 186, 221, 221a, 243, 260, 288, 330
Soft-coated Wheaton terrier: 9a, 10, 22, 42, 81, 159a, 166, 172, 220, 245, 253, 256, 266a, 270, 330
Spinoni Italiani: 91, 103
Standard dachshund: 1, 4, 5, 42, 54, 55, 61, 65, 72, 75, 78, 80, 83, 85, 94a, 103, 115, 129, 135, 156, 161, 166, 173, 176, 179, 181, 187, 199, 214, 216, 220, 224, 228, 230, 237, 239, 245, 250, 256, 267, 275, 296, 297, 326, 328, 330
Standard Manchester terrier: 42, 72, 109, 135, 166, 185, 186, 256, 330
Standard poodle: 9a, 10, 21, 22, 27, 31, 42, 61, 81, 88, 103, 109, 110, 121, 124a, 135, 140, 144, 146, 147, 152, 159a, 166, 175, 184, 186, 199, 220, 221, 221a, 223, 230, 245, 256, 269, 273a, 311a, 330
Standard schnauzer: 23, 42, 62, 121, 147, 166, 221, 221a, 240, 260, 270, 330
Sussex spaniel: 38, 42, 88, 103, 270
Swiss mountain dog: 221, 221a, 249, 311, 311a
Tibertan mastiff: 95, 152, 158, 166, 192, 221, 245, 312, 330
Tibetan terrier: 14, 42, 103, 186, 199, 214, 245, 256, 270
Toy Manchester terrier: 42, 166, 186, 256, 330
Toy poodle: 5, 10, 22, 26, 27, 42, 49, 78, 81, 88, 92, 93, 103, 109, 110, 111, 121, 124a, 135, 136, 140, 144, 146, 147, 149, 156, 165, 166, 173, 175, 184, 193a, 199, 221, 221a, 223, 226, 230, 235, 236, 256, 269, 275, 294, 327, 330

Vizsla: 42, 68, 80, 103, 109, 117, 121, 140, 143a, 147, 152, 161, 166, 193, 221, 221a, 256, 273a, 289, 296, 305, 318, 319, 326

Weimaraner: 31, 43a, 61, 65, 88, 103, 105, 114, 121, 131, 140, 142, 147, 152, 158, 160a, 165, 166, 168, 169, 170, 196, 206, 256, 273a, 289, 296, 305, 311b, 318, 319, 326

Welsh springer spaniel: 42, 135, 152, 245, 256

Welsh terrier: 42, 135, 166, 186, 330

West Highland white terrier: 2, 4, 9a, 10, 21, 22, 42, 68, 71, 81, 106, 136, 156, 159a, 171, 181, 185, 186, 199, 245, 270, 276, 331, 331a

Whippet: 42, 61, 70, 80, 103, 143a, 166, 186, 221, 221a, 234, 256, 330

Wire-haired fox terrier: 3, 22, 42, 78, 87, 88, 103, 112, 135, 138, 185, 186, 243, 256, 260, 288, 310, 330

Wire-haired pointing griffon: 152, 206a, 226

Yorkshire terrier: 42, 71, 80, 88, 103, 149, 154, 162, 166, 179, 181, 185, 235, 236, 245, 256, 269, 270, 276, 330

Section II:

1. Aberrant cilia: eyelashes growing abnormally, such as rubbing against the eyeball (see #88).

2. Abnormal copper metabolism: (usually Bedlington terriers or Doberman pinschers) an inability to utilize and store copper properly, resulting in liver disease and other problems.

3. Abnormal dentition: abnormal placement, number and development of teeth.

4. Acanthosis nigricans: (usually dachshunds) a skin disease where the skin becomes thickened and dark, primarily affecting the axillae (armpits).

5. Achondroplasia: abnormal development of cartilage leading to dwarfism (seen aberrantly in most breeds, but that's what makes a Basset hound and other achondroplastic breeds long and low).

6. Acne: same thing as in humans, affects the muzzle and lip areas.

7. Acral lick dermatitis: a skin disease caused by an animal's licking a localized area excessively, especially on the legs and paws.

8. Acral mutilation: a progressive self-mutilation of the feet and legs. Also known as peripheral sensory neuropathy in pointing breeds because they are born without pain sensation.

9. Acute moist dermatitis: known as "hot spots," a localized area of a severely itchy, inflamed and oozing dermatitis exacerbated by the animal's intense licking and chewing at the spot.

9a. Addison's disease (hypoadrenocorticism): a disease characterized by inadequate secretion of cortisone from the adrenal glands. More common in Old English sheepdogs, standard poodles and bearded collies. (See #159a.)

10. Allergies: same as in humans. Dogs can be allergic to things they come in contact with, eat or inhale.

11. Amyloidosis: a condition where abnormal deposits of proteinaceous material called amyloid are laid down in tissues and impair their function. Common in Akitas and Chinese shar peis.

12. Anasarca: a condition where neonatal puppies have an abnormal accumulation of fluids in their tissues. Often seen in English bulldogs.

13. Anemia with chondrodysplasia: a condition of Alaskan malamutes where there is malformation and maldevelopment of cartilage and red blood cells. Also called stomatocytosis because of the mouth-like shape of red blood cells.

14. Anesthetic idiosyncracy: a condition where an individual has an abnormal response to commonly used anesthetics sometimes leading to death. Idiosyncratic means there is no good explanation or way to predict this.

15. Anomaly of third cervical vertebra: a malformation of one of the neck bones in the spinal column.

16. Anophthalmia: a condition where an animal is born without any eyes.

17. Aortic body tumors: a cancer that arises from a small organ at the base of the aorta, near the heart.

18. Apocrine gland tumor: a cancer arising from glands which secrete fluids (such as mammary glands).

19. Arteriovenous fistula: an abnormal connection that forms between arteries and veins.

20. Ataxia: (see #255, #288.)

21. Atopic dermatitis: a skin disease caused by a dog's reaction to an inhalant allergy. (See #81.)

22. Atopy: an allergy caused from things dogs inhale.

23. Atresia of nasolacrimal puncta: a condition where the holes on the inside of the lower eyelids (puncta) are too small or closed so tears spill over the lid instead of draining to the nose.

24. Atrial septal defects: a malformation of the dividing wall between two chambers of the
heart, usually resulting in a hole which then causes blood flow abnormalities.

25. Avulsion fractures: fractures of the bones caused by a piece being ripped away, usually because of its strong attachment to a ligament or tendon.

26. Basal cell tumor: a cancer arising from a type of skin cell.

27. Behavioral abnormalities: a whole range of abnormal behavior patterns, such as aggression, panic disorders, etc.

28. Bithoracic ectromelia: a condition where the front leg bones are very small or absent.

29. Blepharospasm: an abnormal tightening of the muscles around the eyes, which causes repetitive blinking.

30. Blindness: an inability to see due to a large variety of causes.

31. Bloat: a condition where a dog's stomach produces excessive gas and enlarges severely enough to cause death without immediate treatment. Usually associated with gastric torsion (see #131).

32. Blue eyes: an adverse reaction to certain vaccines containing canine hepatitis virus (adenovirus 1) which produces a bluish discoloration to the cornea.

33. Bullous pemphigoid: an autoimmune (i.e., the body attacking itself) disease associated with the formation of painful blisters.

34. Bundle branch block: an abnormality of the electrical conduction mechanism of the heart.

35. Bundle of His degeneration: a condition where a part of the electrical system of the heart deteriorates.

36. Calcinosis circumscripta: the development of lumps of hard calcium deposits in the skin.

37. Cancer, bladder: as in humans, cancer arising from the bladder.

37a. Cardiac valvular disease: weakness of heart valves producing heart murmurs and leading to cardiac failure. Prevalent in Cavalier King Charles spaniels.

38. Cardiomyopathy: a disease of weakened heart muscles. Common in giant breeds, boxers, Doberman pinschers.

39. Carotid body tumors: cancer arising from a small organ located on the carotid in the neck.

40. Carpal subluxation: a condition where the "wrist" bones are loose and out of alignment.

41. Castration responsive dermatosis: a skin condition characterized by loss of hair, thickened skin and inflammatiion which responds to castration (i.e., hormonally dependent).

42. Cataract: as in humans, a change in structure of the lens of the eye leading to cloudiness and usually to blindness.

43. Cataract with microphthalmia: a condition where a dog has cataracts along with abnormally small eyes.

43a. Cell-mediated immunodeficiency: a deficiency of t-lymphocyte function causing impaired immunity, chronic recurrent infections and stunted growth. Seen in Weimaraners and Dachshunds. (See #311b).

44. Cellulitis (folliculitis and furunculosis): inflammation and infection of the cells of the skin including the hair follicles and deeper structures.

45. Cerebellar cortical abiotrophy: malformation of the neurons in the cerebellum, a part of the brain.

46. Cerebellar degeneration: a condition where a part of the brain deteriorates.

47. Cerebellar and extrapyramidal abiotrophy: a condition where the neurons in the cerebellum part of the brain and parts of the spinal cord are malformed and eventually malfunction.

48. Cerebellar hypoplasia: a condition where the cerebellum, a part of the brain, is poorly formed (too small or absent) and doesn't function properly or at all.

49. Cerebrospinal demyelination: a condition where the neurons of the brain and spinal cord are malformed, lacking a specialized sheath, which causes malfunction.

50. Cervical disc disease: a degeneration or malformation of the cushioning discs between the spinal column bones (vertebrae) in the neck.

51. Cervical vertebral malformation or instability: a malformation of the vertebrae in the neck usually leading to nerve damage. Commonly seen in Doberman pinschers and causes a hackneyed gait (high stepping).

52. Choroidal hypoplasia: improper development of a part of the eye.

53. Ciliary dyskinesia: congenital trait in several breeds where all ciliated cells (those with hairs or moving appendages) are deformed and rigid. Also called immotile cilia syndrome and Kartagener's syndrome. Causes chronic pneumonia and sterility. Common in bichon frise and Doberman pinschers.

54. Cleft lip: a condition where the two halves of the upper lip do not join together. Cleft palate and cleft lip are often seen together.

55. Cleft palate: a condition where the roof of the mouth is not closed and the inside of the nose opens into the mouth.

56. Coliform enteritis: an inflammation of the bowel tract caused by a certain group of bacteria.

57. Collapsed trachea: a condition where the cartilage rings that make up the trachea are malformed and tend to collapse easily.

58. Collie eye anomaly: a condition of collie dogs where the narrow shape of the head causes the eyes to be malformed.

59. Coloboma: an abnormal development of the eye, usually seen in collies, which can lead to blindness (see #203).

60. Colobomas with aphakia: same as above but with congenital absence of the lens.

61. Color mutant alopecia: a condition where certain colored areas of a dog's skin grows less or no fur. Commonly seen in Yorkshire terriers, blue Doberman pinschers and fawn Irish setters.

61a. Combined immunodeficiency: a severe combined deficiency of cell-mediated immunity (T-cell function) and low levels of serum immunoglobulins (IgA, IgG, and variably IgM). Affected puppies usually die from viral infectins by 12-16 weeks of age. Seen in Basset hounds. (See 343a, 168, 169, 170.)

61b. Complement deficiency: a deficiency in serum levels of the third component of complement which impairs neutrophil function and causes recurrent infections. Seen in Brittany spaniels.

62. Conjunctivitis: an inflammation of the conjunctival membrane of the eye.

63. Conus septal defect: a developmental abnormality of the right ventricle of the heart.

64. Copper storage abnormalityin liver: (see #2.)

65. Corneal dystrophy: an abnormality of the cornea usually characterized by shallow pits in the surface.

66. Corneal leukomas: an abnormal accumulation of a white material in or on the cornea.

67. Corneal ulcer, superficial: an erosion of the outer membrane and outer surface of the cornea.

68. Craniomandibular osteopathy: an abnormal development of the bones of the face and the jaw. Seen in West Highland white and Cairn terriers, among other breeds.

69. Cranioschisis: abnormal development of the skull characterized by openings between or in the bones.

70. Cryptorchidism: a condition where one testicle does not descend into the scrotal sac.

71. Cushing's disease (hyperadrenocorticism): a disease characterized by an excess secretion of cortisosteroids from the adrenal glands. (See #156.)

72. Cutaneous asthenia: a condition where the skin lacks its normal strength, elasticity and sensation. Also called Ehlers-Danlos syndrome. Seen in several breeds, including English springer spaniels and boxers. (See #94a.)

72a. Cutaneous vasculopathy: a disease of autosomal recessive inheritance in which young puppies exhibit footpad swelling and depigmentation, with crusting and ulceration of the tips of ears and tail. Seen in German shepherd dogs.

73. Cyclic neutropenia: a condition characterized by periodic lowering of neutrophils, a type of white blood cell. Commonly seen in grey collies.

74. Cystic ovaries: a condition where the ovarian follicles become cyctic (fluid filled) leading to hormonal imbalances and other problems.

75. Cystinuria: an abnormal excretion of a substance (cystine) in the urine.

76. Cystitis and cystic calculi: infection of the bladder which often leads to formation of abnormal mineral deposits (bladder stones).

77. Dacrocystitis: inflammation of a tear sac.

78. Deafness: an inability to hear, due to many different causes.

79. Deformed tail: a congenital condiiton where the tail is malformed.

80. Demodicosis: a kind of skin disease (mange) caused by microscopic Demodex canis mites living within the skin layers and producing an immunodeficiency syndrome.

81. Dermatitis, atopic: inflammation and subsequent infection of the skin due to atopy (see #21 and #22).

82. Dermatomyositis: a disease affecting the skin and muscles, usually in collies or Shetland sheepdogs.

83. Dermoid cyst: a small growth composed of skin-like structures.

84. Dermoid sinus: similar to a dermoid cyst but usually larger. Seen in Rhodesian ridgebacks.

85. Diabetes mellitus: a metabolic disease caused by insulin deficiency and characterized by the inability to utilize sugars normally.

85a. Digital hyperkeratosis: a condition of Irish terrier puppies that causes marked thickening of the foot pads. Affected feet crack, become infected and painful.

86. Discoid lupus erythematosus: a form of autoimmune disease affecting the skin.

87. Dislocation of shoulder: a condition where the bones of the shoulder joint are out of proper position.

88. Distichiasis: abnormally growing eyelashes.

89. Dwarfism: an abnormality of the normal growth pattern resulting in an undersized individual.

90. Dystocia: complications of the birth process (difficult birth).

91. Eclampsia: convulsions usually seen around the time of parturition (whelping).

92. Ectodermal defects: any of a multitude of abnormalities arising from maldevelopment of the fetal ectoderm (e.g., skin, nervous system, eyes).

93. Ectopic ureters: the ureters (tubes leading from the kidneys to the bladder) do not empty into the bladder in the normal location.

94. Ectropion: an abnormal rolling out of the eyelids.

94a. Ehlers-Danlos syndrome: a connective tissue disease characterized by loose, hyperextensible and very fragile skin that tears easily (see #72).

95. Elbow dysplasia: an abnormal development of the elbow joint.

96. Elbow joint malformation: (see #95.)

97. Elbow subluxation: a condition where the elbow joint is loose and out of alignment.

98. Elongated soft palate: the soft palate is abnormally long and causes breathing disorders.

98a. Encephalitis: an inflammatory condition of the brain causing signs of central nervous system dysfunction and epilepsy (seizures). A unique form of encephalitis is prevalent in the pug breed and is called "Pug Dog Encephalitis." (See #109.)

99. Endocardial fibroelastosis: an abnormal condition of scarring of the muscles of the heart.

100. Endometritis: inflammation of the internal layer of the uterus.

101. Enlarged foramen magnum: a condition in which the opening in the skull where the vertebral column begins is too large.

102. Enostosis: a bony growth within the hollow part of a bone.

103. Entropion: an abnormal rolling in of the eyelid.

104. Eosinophilic granuloma: an allergic reactive syndrome characterized by the plaque-like accumulation of eosinophils, a type of white blood cell.

105. Eosinophilic panosteitis: a painful inflammatory bone disease of young, rapidly growing dogs, often characterized by increased eosinophils in the blood. (See #231.)

106. Epidermal dysplasia: abnormal development of the outer layer of the skin. Common in West Highland white terriers and begins in puppyhood. (See #331.)

107. Epidermoid cyst: a small growth consisting of tissues of the outer layer of the skin (see #274.)

108. Epidermolysis bullosa: an abnormal looseness to the skin characterized by large, deep, blister-like lesions.

109. Epilepsy: a disease characterized by convulsions (seizures) and/or disturbances of consciousness.

110. Epiphora: abnormal draining of tears often due to overproduction.

111. Epiphyseal dysplasia: abnormal development of the epiphysis, a part of the long bones.

112. Esophageal achalasia: a functional stricture or spasm of the muscles of the esophagus where it joins the stomach.

113. Esophageal dilatation: an abnormally large and usually flaccid esophagus.

114. Eversion of nictitating membrane: a condition where the third eyelid is protruding.

115. Eye abnormality: any of a number of problems with the eye.

116. Facial fold dermatitis: an infection of the facial skin caused by unusual or excessive skin folds (seen in dogs such as the Pekingese or Chinese shar pei).

117. Facial nerve paralysis: a decrease or cessation of function of the facial nerve leading to a drooping of the affected side of the face.

118. Factor I deficiency or hypofibrinogenemia: a rare deficiency of a clotting factor (fibrinogen), which causes excessive bleeding.

119. Factor II deficiency or hypoprothrombinemia: a rare deficiency of prothrombin, a clotting factor needed to control bleeding.

120. Factor VII deficiency: a mild bleeding disease primarily of beagle dogs.

121. Factor VIII deficiency or hemophilia A: the most common severe inherited clotting disorder of humans and nonhuman animals. Inhereited as a sex-linked recessive trait (carried by females and manifested in males). Affects most dog breeds.

122. Factor IX deficiency or hemophilia B: same as hemophilia A, but more rare and involves a different clotting factor. Affects about 20 dog breeds.

123. Factor X deficiency: a rare clotting disorder primarily of American cocker spaniels. An autosomal trait (affects both sexes).

124. Factor XI deficiency: a rare clotting disorder of several dog breeds. Protracted bleeding from surgical procedures is a feature. Affects both sexes.

124a. Factor XII deficiency: a clotting factor deficiency that rarely produces clinical signs. Seen in toy and standard poodles and occasionally in other breeds. Usually diagnosed incidentally during blood testing for potential bleeding disorder.

124b. Familial amaurotic idiocy: deposits of fatty pigments in the brain produce loss of vision, stuppor, and seizures. Seen in English setters, Germn short-haired pointers, and Australian cattle dogs. (See #177, 193a, 214.)

124c. Fanconi syndrome: a kidney tubular dysfunction of basenjis which leads to glycosuria (see #268).

125. Fibrosarcoma: a cancer arising from certain types of fibrous cells.

126. Fibrous histiocytoma: a type of fibrous tumor arising from cells called histiocytes.

127. Flank sucking: a behavioral problem common in Doberman pinschers and exhibited as a continually wet patch on the flank (from sucking the skin).

128. Fold dermatitis: an inflammation of skin folds especially in dogs with loose skin (e.g., Chinese shar pei).

129. Folliculitis: an infection of the hair follicles.

129a. Fragmented coronoid process: osteochondrosis of the elbow joint. (See #221a.)

130. Furunculosis: an infection of the deeper structures of the skin.

131. Gastric torsion: a condition where the stomach twists, thereby impeding input and output. (See #31.)

132. Generalized myopathy: a condition affecting all the muscles of the body which produces weakness.

133. Genu valgum: malformation of the knee joint ("knock-kneed").

134. Gingival hyperplasia: overgrowth of the gum tissues.

135. Glaucoma: abnormally high pressure in the eye.

136. Globoid cell leukodystrophy: abnormal development and/or function of certain types of white globoid cells in the brain. (See #193a.)

137. Glycogen storage disease: a syndrome characterized by an inability to store and utilize carbohydrates. (See #193a.)

138. Goiter: a swelling of the thyroid gland.

138a. Granulocyte dysfunctin or adhesion defect: an impairment of neutrophil function or adhesion which causes chronic recurring infections, stunted growth and secondary increase in immune globulins (hypergammaglobulinemia). See in Irish setters and Doberman pinschers.

139. Granulomatous colitis: a type of chronic inflammation of the colon characterized by reactive tissue growths.

140. Granulomatous sebaceous adenitis: a disease of sebaceous (sweat) skin glands characterized by reactive tissue growth and autoimmune destruction of the sebaceous glands. Hair loss occurs and is poorly responsive to treatment. Common in standard poodles, Akitas, samoyed and vizslas.

141. Hair follicle tumors: abnormal growths of the hair follicles.

142. Hairlessness: also called alopecia or loss of hair. Can be a normal pattern for breeds like the Mexican hairless dog.

143. Hanging tongue: a syndrome where the tongue does not retract into the mouth properly, due to neurologic or anatomic defects. Commonly seen in Cavalier King Charles spaniels.

143a. Hemangiosarcoma: a cancer of blood vessels involving liver, spleen or skin.

144. Hemeralopia: inability to see in daylight.

144a. Hemorrhagic gastroenteritis: an acute disorder characterized by bloody diarrhea, elevated hematocrit and shock. Common in miniature schnauzers.

145. Hemivertebra: a particular kind of malformation of the vertebra where only half of the structure is formed.

146. Hemolytic anemia: anemia caused by the destruction of the red blood cells by an autoimmune process. Particularly common in cocker spaniels and Old English sheepdogs, as well as several other breeds.

147. Hemophilia A: a blood clotting disorder due to deficiency of coagulation factor VIII (this is the most common type of hemophilia in dogs). (See #121.)

148. Hemophilia B: a blood clotting disorder due to lack of coagulation factor IX. (See #122.)

149. Hepatic portosystemic shunt or arteriovenous fistula: a malformation of blood vessels in the liver or an abnormal communication between the arteries and veins in the liver. 149a. Hepatic lipidosis: an abnormal accumulation of lipids in the liver which leads to liver failure. Common in miniature schnauzers and Shetland sheepdogs.

149b. Hereditary nephritis: also called "Samoyed hereditary glomerulopathy," a sex-linked disease of young males. Affected dogs have renal glomerular disease which rapidly progresses to kidney failure and death. Female carriers have abnormal glomerular basement membrane as well, but usually remain healthy until later life when renal failure may occur.

149c. Hereditary spinal muscular atrophy: an autosomal dominant degenerative disease of motor neurons characterized by weakness and muscle atrophy with a typical gait, and progressing to dangling of the head and a drooping, paralyzed tail. Severely affected dogs become paralyzed and die by 3-4 months of age. Seen in Brittany spaniels.

149d. Hepatocerebellar degeneration: a syndrome of progressive cerebellar and hepatic disease of 6-8 week old Bernese mountain dogs with lesions of cerebellar abiotrophy and coexistant hepatic lesions. Autosomal recessive inheritance.

150. Hermaphroditism: a syndrome where the individual has anatomical features of both sexes.

151. Heterochromia, iris: the presence of different colors in the same or both irises.

152. Hip dysplasia: a developmental malformation or subluxation of the hip joints.

153. Histiocytoma: a tumor composed of certain skin tissue cells (i.e., histiocytes).

154. Hydrocephalus: a condition where there is an abnormal accumulation of fluid in the ventricles of the brain.

155. Hygroma: a fluid-filled sac usually occurring on the elbows of large breed dogs such as the Great Dane or Irish wolfhound.

156. Hyperadrenocorticism or Cushing's disease: a disease where the adrenal glands are overactive. (See #71.)

157. Hypercholesterolemia: a disease where the animal has too much cholesterol in the blood system. Commonly associated with hypothyroidism.

158. Hypertrophic osteodystrophy: a condition of rapidly growing giant breeds where there is an abnormal inflammation of bones with pain and development of excessive bony growths.

159. Hypertrophy of nictitans gland: a condition where the gland of the third eyelid is abnormally large.

159a. Hypoadrenocorticism: a disease where autoimmune or other causes of destruction of the adrenal glands produces a deficiency of corticosteroids. (See #9a.)

160. Hypoglycemia: a syndrome where the animal has an abnormally low blood glucose.

160a. Hypomyelinogenesis: failure of the nervous system to form myelin, seen at birth.

161. Hypopigmentation, lips and nose: a condition where an animal lacks pigment (color) in areas where it is usually present. (See #328.)

162. Hypoplasia of dens: a condition where part of the second vetebra fails to develop fully and leads to instability.

163. Hypoplasia of larynx: a condition where the larynx (cartilage of the "voice box") fails to develop fully.

164. Hypoplasia of trachea: a trachea that fails to develop fully.

165. Hyposomatotropism: failure of the body growth hormones (somatomedins) to develop fully. Also known as growth hormone-responsive dermatosis. Common in Pomeranians. 166. Hypothyroidism: a common endocrine disease where the body produces an abnormally low amount of thyroid hormones. An autoimmune destruction of the thyroid gland which affects more than 50 dog breeds. (See #192, #312.)

167. Hypotrichosis: a condition where there is an abnormally small amount of hair growth.

168. Immunoglobulin A deficiency: a condition where levels of secretory immune globulins are low. Common in Chinese shar peis and beagles. (See #187.)

169. Immunoglobulin G deficiency: a condition where circulating antibody levels are low. Produces immune deficiency and susceptibility to infections.

170. Immunoglobulin M deficiency: a condition where antibodies produced in early stages of an immune response are low, producing susceptibility to infection. Seen in Doberman pinschers.

171. Inguinal hernia: a break in the muscular layer of the body wall occurring at the inguinal canal (where the back leg meets the body).

172. Intestinal malabsorption: a disease where the intestinal tract does not absorb nutrients properly. Also known as protein-losing enteropathy as a consequence of inflamatory bowel disease. In Irish setters there is also a wheat-sensitive enteropathy.

173. Intervertebral disc disease: a disease where the discs between the vetebra are abnormal and prone to rupture and misplacement.

174. Intussusception: a serious condition where the intestinal tract telescopes in on itself.

175. Iris atrophy: a condition where the iris (the colored part of the eye) shrinks and becomes non-functional.

176. Iris heterochromia: a condition where one iris is a different color from the other or has more than one color to it.

177. Juvenile amaurotic idiocy: a syndrome characterized by early onset blindness and low mental capacity.

178. Juvenile cellulitis: an inflammation of cells (usually skin cells) occurring in the young animal.

179. Keratitis sicca: a condition where one or both eyes do not produce a normal amount or type of tears.

180. Keratoacanthoma: a small growth, usually on the face, filled with keratin material.

181. Keratoconjunctivitis sicca: (see #179.)

182. Kidney aplasia, unilateral: a developmental abnormality where one kidney fails to develop. Also called renal agenesis.

183. Kinked tail: a developmental abnormality where the tail has a pronounced kink.

184. Lacrimal duct atresia: a condition where the duct draining tears from the eye is too small or not formed.

185. Legg-Perthes disease: a disease where the blood vessels feeding the femoral head (top part of the thigh bone) shrink, leading to starvation and death of the femoral head (the ball of the ball-and-socket joint of the hip). Also called Legg-Calve'-Perthes disease.

186. Lens luxation: a condition where the lens in the eye is displaced into an abnormal position.

187. Linear IgA dermatosis: a type of skin disease resulting from an abnormality of the secretory immune system. Common in Chinese shar peis.

188. Lip fold dermatitis: a skin infection caused by redundant skin folds around the mouth.

188a. Lipidosis: a form of lysosomal storage disease where lipids accumulate in nerves. Called GM-1 gangliosidosis in Portuguese water dogs. (See #193a.)

189. Lissencephaly: an abnormal brain development where the surface lacks gyri (the grooves).

190. Lung torsion: a condition where one or more lung lobes twist upon themselves.

191. Lymphedema: a disorder where valvular blockage of lymph flow or twisted lymphatic ducts causes an acumulation of fluid to swell tissues with edema.

192. Lymphocytic thyroiditis: an autoimmune disease causing inflammation and destruction of the thyroid gland, which becomes infiltrated with lymphocytes (white blood cells) and leads to hypothyroidism. This is the most comon endocrine disease of the dog and has an inherited predisposition (see #166, #312).

193. Lymphosarcoma: a cancerous condition involving the lymphatic system. One of the more common canine cancers.

193a. Lysosomal storage diseases: a group of progressive multifocal neurologic disorders caused by specific enzyme deficiencies leading to death of nerve cells and accumulation of their respective enzyme substrates in cells. (See #299a.)

194. Malabsorption syndrome: (see #172.)

195. Malocclusion: a condition where the teeth do not meet properly.

196. Mastocytoma: a rare cancer developing from a type of tissue cell known as a mast cell.

197. Melanoma: a rare cancer developing from the type of skin cell which produces pigment (melanin).

198. Metabolic bone disease: any of a number of diseases affecting the bones due to an abnormality of metabolism.

199. Microphthalmia: a condition where one or both eyes are too small.

200. Missing teeth: a condition where there are too few teeth.

201. Mitral valve defects: a group of abnormalities of the mitral valve of the heart.

202. Mononephrosis: a condition where only one kidney is present.

203. Multiple colobomas: a developmental abnormality of the structures of the eye.

204. Multiple epiphyseal dysplasia: a condition where many of the long bones develop abnormally due to changes in the growth plates.

204a. Muscular dystrophy: a congenital and often inherited form of generalized muscle dysfunction which causes signs such as poor growth, weakness, abnormal gait, dificulty eating and swallowing, and muscle atrophy. Affected animals have serious health problems and may die or be euthanized. Inheritance is sex-linked in Golden retrievers, Irish terriers, Samoyeds, and Belgian shepherds.

205. Muzzle pyoderma: an infectious skin disease on the muzzle of an animal.

206. Myasthenia gravis: a syndrome characterized by muscle fatigue due to an autoimmune disease which produces chemical abnormalities of the muscles and nerves. An enlarged esophagus called megaesophagus can result and causes regurgitation of food.

206a. Narcolepsy: a neurological disorder characterized by falling asleep suddenly (collapse) which can occur during periods of activity and last for varying lengths of time. Seen in Doberman pinschers and Labrador retrievers.

207. Narrow palpebral fissure: an abnormally small opening between the upper and lower eyelids.

208. Nasal pyoderma: a skin infection of the nose.

209. Nasal solar dermatitis: a skin disease of the nose and muzzle which is greatly affected by exposure to sunlight. Common in collies.

210. Nasolacrimal puncta atresia: (see #23.)

211. Necrotizing myelopathy: a condition where the spinal cord gradually dies.

212. Necrotizing panotitis: a severe infection of the ear and surrounding tissues.

213. Neuromuscular atrophy: a condition where the muscles waste away due to lack of proper nerve supply.

214. Neuronal ceroid lipofuscinosis: a congenital disease where fatty pigments are deposited in the brain and cause brain dysfunction. (See #193a.)

215. Neurotropic osteopathy: a disease of the bones due to abnormalities of the nerves.

216. Nodular panniculitis: a skin disease characterized by nodules of inflammation under the skin.

217. Oligodendroglioma: a cancer arising from a type of cell found in the brain and spinal cord.

218. Oligodontia: an abnormally small number of teeth.

219. Open fontanel: a condition where the suture lines between bones of the skull do not fuse together properly.

220. Optic nerve hypoplasia: a condition where the optic nerve going from the eye to the brain is too small.

221. Osteochondritis dissecans: a specific form of inflammation of the cartilage of certain joints which causes arthritis. (See #221a.)

221a. Osteochondrosis: a group of developmental diseases resulting in abnormal formulation of joint cartilage. Commonly involves the shoulder, stifle, hock or elbow. (See #221.)

222. Osteodystrophy: any of a number of diseases involving the development of the bones.

223. Osteogenesis imperfecta: imperfect development of the structure and/or mineralization of the bones.

224. Osteopetrosis: a condition where the bones are abnormally dense and hard.

225. Osteosarcoma: a cancer arising from the cells of the bones.

226. Otitis externa: an infection of the external structures of the ear.

227. Otocephalic syndrome: a developmental abnormality where the animal lacks a lower jaw, and the ears meet below the face.

228. Overshot jaw: a condition where the upper jaw is too long for the lower jaw.

229. Pancreatic insufficiency: a condition where the pancreas does not produce the proper enzymes for digesting food.

230. Pannus: an immunologic eye disease characterized by abnormal growth of tissue over the cornea.

231. Panosteitis: (see #105.)

232. Parosteitis: inflammation of tissue around a bone.

233. Parotitis: inflammation of the parotid salivary gland. Also called parotiditis.

234. Partial alopecia: some loss of the normal haircoat.

235. Patella luxation: a condition where the knee caps slide in and out of place.

236. Patent ductus arteriosus: failure of the vessel remnant joining the aorta and pulmonary artery in fetal life to close properly at birth, thereby shunting blood away from the lungs.

237. Pattern alopecia or baldness: hair loss occurring in certain patterns. Common in dachshunds.

238. Pemphigus erythematosus: one of many skin diseases caused by an autoimmune mechanism.

239. Pemphigus foliaceous: another skin disease caused by autoimmune destruction of tissues.

240. Perianal adenoma: a cancer arising from a cell of a gland found near the anus.

241. Perianal fistulas: a condition characterized by abnormal communications from deeper tissues to the skin surrounding the anus.

242. Perianal gland tumor: (see #240.)

243. Persistent right aortic arch: a developmental abnormalitiy where one of the fetal blood vessels near the heart does not atrophy as it should.

244. Persistent hyaloid artery: as #243, however, involving a blood vessel inside the eye.

245. Persistent pupillary membrane: a developmental abnormality where the membrane forming the iris does not form properly.

245a. Phosphofructokinase deficiency: a deficiency of a specific red blood cell enzyme in English springer spaniels. Causes chronic anemia, exercise-induced acute hemolytic crises and enlarged spleen.

246. Pigmentary keratitis: an inflammatory condition of the cornea characterized by abnormal pigmentation.

247. Pituitary dwarfism: a developmental abnormality resulting in an undersized animal due to a defective pituitary gland.

248. Pituitary tumor: a cancer arising from the pituitary gland.

249. Platelet disorder: a group of abnormalities of small blood cells necessary to control bleeding. (See #311.)

250. Pododermatitis: a skin infection of the paws.

251. Polyostotic fibrous dysplasia: a type of bone disease where the bones are composed of improper fibrous tissues.

252. Polyradiculoneuritis: an acute inflammatory disease of several groups of nerves causing fever.

253. Posterior retinal atrophy: a deterioration of the part of the eye which translates light to electric impulses (the retina). Produces night blindness. (See #256.)

254. Primary peripheral retinal dystrophy: a certain type of developmental disease affecting the retina.

255. Progressive ataxia: a condition where the animal's sense of coordination deteriorates.

256. Progressive retinal atrophy: a disease where the retina slowly deteriorates, producing night blindness.

257. Prolapsed rectum: a condition where the inside of the rectum protrudes outside the anus.

258. Prolapsed uterus: a condition where the uterus protrudes into the vaginal canal or through the vaginal opening.

259. Pseudohermaphrodism (pseudohermaphroditism): a condition where the animal has the gonads of one sex but the appearance is ambiguous or is of the opposite sex.

260. Pulmonic stenosis: a condition where one of the valves of the heart does not open properly.

261. Pyloric stenosis: a condition where the opening leading from the stomach does not function properly.

262. Pyometra: an bacterial infection of the uterus where it fills with pus.

263. Pyruvate kinase deficiency: a deficiency of a specific red blood cell enzyme. Most commonly seen in basenjis; also in beagles and Cairn terriers.

264. Quadriplegia with amblyopia: a syndrome characterized by weakness of all four limbs, as well as of vision.

264a. Rage syndrome: sudden unprovoked aggression of serious nature. Seen in English springer spaniels. (See #27.)

265. Recessive retinal dysplasia: a developmental disorder resulting in an abnormal retina, carried by a recessive gene.

266. Renal cortical hypoplasia: a condition where the cortex of the kidney(s) develops incompletely.

266a. Renal dysplasia: a condition where the kidneys form abnormally. Renal failure develops with protein loss in urine.

267. Renal hypoplasia: a condition where the kidney(s) do not develop completely.

268. Renal tubular dysfunction: a condition where the tubules of the kidneys (the filtering structures) do not function properly. (See #124b.) In basenjis, glycosuria develops and is called Fanconi syndrome.

269. Retinal detachment: where the retina is unattached to the back of the eye.

270. Retinal dysplasia: a condition where the retina is malformed.

271. Schnauzer comedo syndrome: a skin disease of schnauzers where the skin forms comedones ("blackheads").

272. Scotty cramp: a condition fround in Scottish terriers where the animal has periodic, generalized cramping of the muscles.

273. Screw tail: a birth defect where the tail is twisted tightly on itself.

273a. Sebaceous adenitis: (see #140.)

274. Sebaceous cyst: a small mass in the skin with a secretory lining and filled with a yellow waxy-like material (see #107).

275. Sebaceous gland tumor: a tumor arising from sebaceous glands of the skin.

276. Seborrhea: a skin disease with excess scaling of the skin and often an excess of sebum (oil-like substance) and odor.

277. Sertoli cell tumor: a tumor of the testicles which secretes estrogen and causes feminization.

278. Short skull: a skull that is abnormally short for the breed in question.

279. Short spine: a spine that is abnormally short for the breed in question.

280. Short tail: a tail that is abnormally short for the breed in question.

281. Shoulder abnormalities: a group of disorders of the shoulder joint due to malformation or subluxation.

282. Shoulder dysplasia: a looseness of the shoulder joint.

283. Silica uroliths: stones which are composed primarily from silicone that form in the bladder.

284. Sinoatrial syncope: a condition where the electrical impulses of the heart are abnormal and the animal has episodes of syncope (fainting).

285. Skin disorders: any of a number of abnormalities of the skin.

286. Skin neoplasms: any number of tumors arising from cells of the skin.

287. Spina bifida: a developmental abnormality where some vetebra are malformed thereby exposing the spinal cord.

288. Spinal cord demyelination (ataxia): an abnormality of the nervous tissue of the spinal cord leading to incoordination.

289. Spinal dysraphism: a developmental abnormality where the spinal cord does not form completely.

290. Spinal osteochondrosis: a specific type of developmental abnormality of the vertebrae.

291. Splenic torsion: a condition where the spleen twists upon itself.

292. Spondylolisthesis (Wobbler's syndrome): a condition where the vertebrae of the neck slip out of joint and are malformed causing progressive incoordination of the rear legs. Commonly seen in Doberman pinschers. (See #51.)

293. Spondylosis: a malformation of the vertebrae.

294. Squamous cell carcinoma: a cancer arising from the squamous type of skin cell.

295. Stenotic nares: a condition where the openings of the nose (nares) are too small.

296. Sterile pyogranuloma syndrome: a disease of the deeper layers of the skin characterized by formation of abnormal tissues, with no infectious organisims involved.

297. Sternal callus: a thickened, hairless area forming on the chest of an animal.

298. Stockard's paralysis: a degeneration of parts of the spinal cord causing paralysis.

299. Stomach torsion: (see #131.)

299a. Storage disease: (see #193a and #188a.)

300. Subaortic stenosis: a tightening of the outflow opening for blood to go from the heart into the aorta. Common in golden retrievers and Newfoundlands.

301. Subcorneal pustular dermatosis: a skin inflammation occurring between certain layers of the skin.

302. Subcutaneous cysts: small fluid-filled masses accumulating under the skin.

303. Subvalvular aortic stenosis: as #300, but the tightening occurs below the aortic valve.

304. Swimmer puppies: a developmental defect which causes a flattening of the body so that newborn pups are unable to place their feet under them for proper locomotion.

304a. Syncope: a brief period of fainting or collapse.

305. Syringomyelia: developmental abnormalities causing cavities within the spinal cord, probably just an effect of #289. Common in Rhodesian ridgebacks.

306. Systemic lupus erythematosus: an autoimmune disease where antibodies form against the nuclear protein of cells. Characterized by skin lesions as well as other organ dysfunctions and blood abnormalities.

307. Tail abnormalities: any number of problems associated with the tail.

308. Tail fold dermatitis: a skin infection caused by abnormal tissue folds around the tail.

309. Teeth abnormalities: any number of problems of the teeth.

310. Tetralogy of Fallot: a specific four-way developmental abnormality of the structures of the heart and associated great vessels.

311. Thrombocytopathy: a functional abnormality of small blood cells (thrombocytes or platelets) which are needed to control bleeding. (See #249.)

311a. Thrombocytopenia: a reduced number of platelets in the blood which causes pinpoint hemorrhages in the skin and mucosa. Often accompanies #146 as an autoimmune syndrome called Evans syndrome. (See #249.)

311b. Thymic atrophy: a deficiency of cell-mediated immunity expressed by decreased T-cell function and low levels of growth hormone. Occurs in Weimaraners. (See #165.)

312. Thyroiditis: an autoimmune inflammatory disease of the thyroid gland. (See #166 and 192.)

313. Tracheal collapse: (see #57.)

314. Trembling of the hindquarters: a condition where the rear legs tremble due to muscle weakness or other pathologies.

315. Type II muscle fiber deficiency: a deficiency in form and/or function of a specific type of muscle fiber.

316. Ulcerative colitis: an autoimmune inflammation of the lining of the colon characterized by formation of ulcers.

317. Ulcerative keratitis: an inflammation of the cornea characterized by the formation of ulcers.

318. Umbilical hernia: a break in the abdominal muscle wall at the point where the umbilical cord enters the body.

319. Undershot jaw: a condition where the lower jaw is too long for the upper jaw.

320. Ununited anconeal process: a developmental abnormality of one of the bones of the elbow joint causing pain. (See #221a.)

321. Uric acid calculi: bladder stones which are formed primarily from urates. Common in Dalmatians.

322. Uric acid excretion abnormalities: an abnormality in the process of the excretion of the uric acid formed during metabolism. Common in Dalmatians.

323. Uterine eclampsia: (see #91.)

324. Uterine inertia, primary: a condition where the uterus does not have the muscular strength to proceed with the birth process, and not due to any acquired problems (e.g., malnutrition).

325. Vaginal hyperplasia: an overgrowth of tissues of the vagina.

326. Vasculitis: an inflammatory condition of the blood vessels.

327. Ventricular septal defect: an abnormality (usually a hole) in the wall between the two chambers of the heart.

327a. Vitamin B12-responsive malabsorption: a disease of young Giant schnauzers in which there is selective inability to absorb vitamin B12 from the bowel. Affected puppies have chronic nonregenerative anemia, low white blood cell counts, low serum vitamin B12, metabolites (methylmalonic acid) in the urine, and failure to thrive.

328. Vitiligo: a lack of pigment in the skin (called vitiligo in man and hypopigmentation in nonhuman animals). Common in rottweilers, Doberman pinschers, Old English sheepdogs and dachshunds. (See #161.)

329. Vogt-Koyanagi-Harada-like syndrome: an autoimmune disease common in Akitas and the "sled" dog breeds where the eyes, blood and other tissues are progressively destroyed leading to blindness and death.

330. von Willebrand's disease: a type of bleeding disorder caused by defective blood platelet function. Occurs in 59 dog breeds but most often in Doberman pinschers. An autosomal trait affecting both sexes.

331. Westie armadillo syndrome: a condition of West Highland white terriers where the skin becomes very thickened. Related to atopic (inhalant) allergies. (See #106.)

331a. White dog shaker syndrome: a disorder mainly of white dogs having muscular tremors over entire body, incoordination and rapid eye movements. Episodes occur with stress or excitement.

332. Wobbler's syndrome: (see #51 and #292.)

333. Zinc deficiency: can be caused by dietary problems, but also from an inability to utilize and store zinc properly. Seen as a lethal problem called acrodermatitis in bull terriers.

334. Zinc-responsive dermatosis: a condition where the skin is abnormal (scaly, hair loss, etc.) but which responds to the administration of zinc in the diet.

10. Readings/Study

There is no other direct literature on derailed evolution. However, Genetic Catastrophe Group (a Yahoo! group) and Derailed Evolution/Genetic Catastrophe (a Google group) are good places to start the study of gene/genetic damage and derailed evolution. Ronald A. Fisher and William D. Hamilton (World's two best mathematical geneticists) understood these things very well, but they did not discover that pollutants that are mutagenic and/or carsinogenic are the real problem.

978-0-595-37353-6
0-595-37353-4